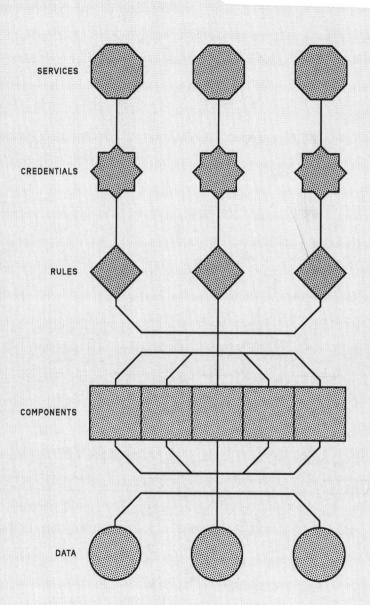

SERVICES

CREDENTIALS

RULES

COMPONENTS

DATA

Praise for *Platformland: An Anatomy of Next-Generation Public Services*

"An updated and nuanced homage to O'Reilly's seminal work on 'government as platform', Richard Pope's *Platformland* adds two necessary elements. First, a very readable near-history of the digitalization of UK public services. Second, a walk-through of the potential of methods and technologies within our reach to fundamentally transform the experience of public services for citizens and those who serve them. For decision makers the question should not be 'Have you read *Platformland*?': it should be 'Why haven't you read it yet?'" — **Theo Blackwell MBE**, Chief Digital Officer for London

"Clear, concise and full of solid examples – this book will really help policymakers understand how to improve the use of technology in public services. Its value can be summed up in one of the chapter headings: 'Complexity simplified'. No prior specialist knowledge is needed to understand the wealth of good advice in this book."
— **Richard Allan**, Baron Allan of Hallam

"Anyone involved in the reform of public services must read *Platformland*. Must, not should. This extraordinary book represents a quantum leap in thinking around public service reform, written from the perspective of a uniquely experienced practitioner. It explains clearly, with examples, how to radically reinvent public services to make the most of the potential of the digital age, while mitigating its downsides. I cannot recommend it highly enough."
— **Tom Loosemore**, Partner, Public Digital

"Richard Pope – one of the most creative thinkers in public sector digital design – shows in this excellent book how the digital revolution in public services is still far from realizing its full potential. *Platformland* fizzes with ideas about how the next generation of digital public services can avoid the pitfalls seen in current systems and make government work better in our everyday lives."
— **Joe Tomlinson**, Professor of Public Law and Director of Administrative Fairness Lab, University of York

"This book is my go-to for 'read this and then we'll talk' in terms of digital government and the future. This is now the beacon in the various futures that could be; we should all be reading it and engaging on which of those futures we choose. It was when I started highlighting too many sentences in quick succession that I realized how packed and efficient the book is at delivering the exact value that anyone interested in the next generation of public services and

infrastructure should be seeking. The book emphasizes ensuring a caring, deliberative and inclusive future in delivering services to people, and it provides detailed and practical methods and frameworks (or 'patterns') to implement our collective humanized digital future."
— **Richard Gevers**, Founder, Open Cities Lab

"Richard Pope presents the most original, persuasive and practical set of design principles for digital public services since the UK Government Digital Service swept the global stage more than a decade ago." — **Richard Sargeant**, Partner, Boston Consulting Group

"The field of digital transformation and the role of platforms in government and public services is only growing – in both opportunity and challenge. Richard Pope has been at the forefront of seizing this opportunity and solving problems, and this book is a testament to his experience and clear-sighted vision for digital transformation that delivers in the public interest."
— **Emrys Schoemaker**, Caribou Digital/London School of Economics

"A resourceful, well-evidenced and easy-to-read guide applying state-of-the-art design thinking to digital public service, based on a deep knowledge of the technically possible, and all without ever losing sight of the 'public' bit: that citizens should expect simplicity, efficiency and empathy in their interactions with government."
— **Morgan Currie**, Senior Lecturer in Data and Society, University of Edinburgh

"Communicating the technical innards of digital public service delivery is really hard, so what Richard Pope has achieved in *Platformland* is quite simply extraordinary. This beautifully written book is a lyrical call to restore humanity to the centre of public service design and delivery. It is a call that must be heeded by governments if we are to solve some of the huge societal issues we face. This is not a book about technology, because in Pope's own words: 'If code is law, then software, the manifestation of code in the world, is politics.'"
— **Emer Coleman**, former Deputy Director, Government Digital Service, and architect of the London Datastore

"*Platformland* achieves the rare feat of providing in-depth analysis of the innards of public sector digitalization programmes while remaining accessible to those without the technical expertise of the author. The book will change the way you think about the design and delivery of public sector digital services. It is a 'must-read' for those tasked with designing, studying and researching these systems."
— **Jed Meers**, Senior Lecturer in Law, York Law School, University of York

Platformland

PERSPECTIVES ON BUSINESS

Series editor: Professor Diane Coyle

Platformland

An Anatomy of Next-Generation Public Services

Richard Pope

LONDON PUBLISHING PARTNERSHIP

Published by London Publishing Partnership
www.londonpublishingpartnership.co.uk

Published in association with
Enlightenment Economics
www.enlightenmenteconomics.com

ISBN: 978-1-916749-11-5 (pbk)
ISBN: 978-1-916749-12-2 (iPDF)
ISBN: 978-1-916749-13-9 (epub)

A catalogue record for this book is
available from the British Library

This book has been composed in Candara

Copy-edited and typeset by
T&T Productions Ltd, London
www.tandtproductions.com

All illustrations created by David Marques
https://davidmarqu.es

Printed and bound in Great Britain
by Hobbs the Printers Ltd

www.carbonbalancedprint.com
CBP2250

Contents

Preface

Waiting. My first memories of bureaucracy are of waiting. Waiting in the car while my mum collected the Family Allowance from the Post Office near school each month. Waiting too at the big Post Office in the town centre, filing glacially past the green and brown forms for driving licences and passports, all stacked up in plexiglass dispensers. Adverts for government schemes and local businesses looping on TVs encapsulated in beige plastic at the hairpins of the queue. Those TVs-that-were-not-TVs seemed like the future in a world with few screens. There was waiting at the building society* too, against the inexplicably carpeted walls, while forms were stamped and scribbled on with pens on chains, and cardboard account books with narrow electromagnetic strips were passed back and forth.

The Family Allowance was part of Britain's post-war social settlement. A universal benefit paid directly to women with children – part of Beveridge's 'abolition of want'. By the 1980s it had long since been renamed Child Benefit, but names stick strongly to the surfaces of public services. Its introduction followed a twenty-five-year campaign by Eleanor Rathbone, a former president of the National Union of Women's Suffrage Societies who was in her final years as a member of parliament by the time her campaign was won in 1945. She remained an MP just long enough to see off an attempt to have the allowance paid directly to men.[1]

---

* In the United Kingdom and in some other countries a 'building society' is a financial organization providing mortgages and other financial services that is owned by its members.

Family Allowance payments were made via Girobank: a fully computerized bank (probably the first in the world) launched by the British government in the 1960s.[2] Girobank's technology was built by the Post Office, which had built the world's first programmable digital computer only a generation before. The whole post-war welfare state was, in implementation, a technology effort. Computing resources, huge for the time, made sure welfare payments were calculated and pensions paid.[3]

Post-offices, paper forms, mainframes – public services are built using the materials and infrastructure of the day. They don't just happen to exist, they are the product of political decisions and, together, they exert an organizing pressure on people's lives. So, as my mum collected her universal income from a futuristic fully automated publicly owned financial institution, my brother and I sat in the car and waited.

In 1980s Britain, the public sector existed in different places and presented a different surface to the world than today. It was not just the red phone boxes and the blue-grey British Rail trains either: the little white vinyl-seated Mini we waited in was probably built by the state as well. It too was once a vision of the future. We were taught to program on BBC Micro computers, seemingly made from the same beige plastic that housed the TVs in the Post Office. School buildings were immediately recognizable, built as they were using a modular system created by a group of local councils. Eating lunch at school, our cutlery was stamped with SCC, for Surrey County Council, bulk ordered from Sheffield foundries.

The shops down the road from school had glazed tiles with the initials of London County Council above their windows. The LCC had brought land outside Greater London after the war as part of the reconstruction effort (you could tell when you were in London proper because the roadside railings were painted municipal blue). Those glazed tiles still appear all over London: part of a policy of mass home building that declared in public space who to hold accountable.[4]

It's not so much that the past is a different country, more that the past was a different future. It turned out that the future did have *a lot* more screens in it (but less carpet on the walls).

*

This is a book about the feel, materials and accountability of the next generation of public services.

We wait differently now. Rather than saving us from bureaucracy, digital has displaced and atomized it. Paper forms have become web forms and countless web pages explaining how to complete them. We wait for emails, rather than letters. Communal experiences with public services, however mundane, have become millions of individual interactions. Government has been placed at a distance – harder to understand and less willing to account for itself when accessed through digital interfaces.

Despite the huge potential for technology and design to improve public services and enrich democracy – the things that got me excited when I stumbled across the emergent civic tech communities of the early 2000s and that eventually led me to working at the Government Digital Service – we have lost something along the way.

I've written *Platformland* in the hope that people can once again get excited about the possibilities of technology, design and democracy. I know I still am.

Introduction

The aim of most digital programmes is the status quo, delivered more cheaply. The digitization of public services has, by and large, happened on governments' terms rather than the public's. Government officials have strong incentives to fund work that cuts the cost of existing services, and creating new ones is too complex. But the things that are expensive and complex for government are not the things that are expensive and complex for the public. Digital rarely gets applied to bureaucratic frictions, the voids that exist between one public service and another, or the ragged edges between the public sector and the rest of everyday life. The public are engaged as consumers, not as part of a democratic society. In a democracy, understanding the way things are is a precondition for changing them. But public sector design optimizes for efficiency and utility. Government is designed out of the way, and too much value is placed on Silicon Valley minimalism, in a way that misunderstands the nature of what makes public services *public*.

This book describes services that work much harder for the public, are simpler to build and fulfil their duty to be understandable, accountable and democratic. For the first time, the systematic elimination of administrative burden across society is within reach. If we choose to follow this path, it will become the expectation that public services automate away as much faff and admin as possible. New public infrastructure will mean it also becomes the expectation within government that creating and changing services is simple, rather than complex. The public will be able to understand how public services work in ways that were previously unthinkable. Those services will be operated transparently, will be trusted to admit their flaws,[1] and will always offer clear opportunities for society to improve them. The relationship between citizen and state will be properly acknowledged and cultivated. It will not be based on a consumer–service provider relationship, but will instead recognize the public as co-producers of real-world outcomes, with empathetic interactions (in both directions) between the public and government officials.

The problem with the future – as Douglas Coupland points out in *Kitten Clone*, his book about internet infrastructure giant Alcatel-Lucent (a 'company you've most likely never heard of' that, nonetheless, we are utterly reliant on) – is that it is difficult to predict because changes tend towards the subtle. 'Will the future be noisier or quieter?' for example.[2] This book tries to equip you to answer questions such as: Will the forms get longer or shorter? Will the relationship with the state get closer or further away? Will we understand more or less about how government works? Will there be more or less certainty in our interactions? More 'computer says no' or more 'computer says maybe'?

Creating the next generation of public services demands both an understanding of the materials they will be made from and a new set of values for public sector design. Seeing the future of public services is not that hard because so much of it is already here. To look to the future, we need only find things that are well-known in digital but have not yet been applied to the public sphere, and then we must look for things from the public sphere that are significant gaps in digital practice. As such, I have tried first to surface a few foundational ideas in computer science, design and digital culture and then to remix those ideas with public policy. I hope that this will help more policy professionals think like technologists and designers, and help more technologists and designers think about public policy and governance.

This book is titled as an 'anatomy' not as a biological metaphor but as the act of separating and dividing something into parts for detailed examination.[3]

Despite the public sector's long history in technology, technology is still somehow perceived as new and novel, talked about in euphemisms and metaphors rarely used by practitioners (portals, data lakes, catalogues, data leaks, hubs, digital twins). But as Jonathan Meades put it: if something is always *like* something else, it is never truly itself.[4] Partly, that's a result of the othering of digital by the classically trained policy class. Partly it's the soothing assertions from management consultants that digital transformation is not about the technology. But computers change how we perceive problems and they give us entirely different ways of thinking.[5] To 'other' digital technology is to cut off that understanding. It's hard to appreciate the true nature, risks and opportunities of data, for example, without experiencing the feel of data in a database; the potential for its tables to join together that tugs at the edge of your mind; the intrinsic understanding of what is quick, what is messy and what is risky. If your mental model is a catalogue, or a filing system or a lake, then those things are lost, or at least diluted.

This book attempts to reset the idea that technology should be perceived as new and novel. Instead it describes the organizational units of digital-age public services on their own terms. As far as possible I have tried to describe digital on its own terms and to stick to terms with established meaning by digital practitioners (platform, bug, etc.).

The ideas in this book draw extensively on the experience of the UK Government Digital Service, and on the design of GOV.UK and Universal Credit in particular. Both of those projects showed the potential for digital to create very different ways of interacting with public services. Neither is without flaws, but it is those flaws that the ideas in this book set out to address. The book also builds on the work of creating a vision for 'Government as a platform' in the UK, which is in turn based on the earlier work of Tim O'Reilly.[6]

Finally, I also discuss examples of forward-looking services and infrastructure from around the world – particularly India and Estonia – collected while writing the Platformland newsletter and from my time spent at the Harvard Kennedy School researching the emergence of public sector infrastructure.

In each chapter you'll find a series of 'design patterns'. A design pattern is a reusable solution to a common problem. Ones here include 'Components support multiple services', 'Proactive services' and 'Publish rules as code'. Each chapter also contains a guiding strategy aimed at realizing the chapter's theme. Examples include 'Digitize all credentials', 'Decouple data from services' and 'Put transparency at the point of use'. You can think of these as 'if you do one thing, do this!' (or maybe as 'credible things to put in a last-minute PowerPoint presentation'). Concluding each chapter is a list summarizing its main points. These boxes can serve as a refresher or a shortcut.

The following terms are used throughout the book.

- *Services* are the things that the public interact with to get an outcome, such as becoming a foster carer or getting a vaccination. They are often a mix of digital and real-world interactions.

- *Credentials* provide digital proof that someone has done something. For example, a successful application for welfare creates a credential someone can use to prove they are receiving it when applying for additional support.

- *Rules* are policy and legislation implemented as software code and available via an API (application programming interface; a way for one computer program to talk to another).

- *Components* are the common bits that make services work, such as sending a message, booking a delivery or verifying a credential.

- *Data* is the foundational information about things such as people, places and organizations, managed as a common resource.

Sometimes, rules, components and data are described collectively as *digital infrastructure.*

You don't need to read this book in order. Each chapter should make sense on its own, and the design patterns and guiding strategies have been written to stand alone too. So, if you are a visual thinker, or have a brain that prefers to jump about between things, then it should work for you as much as a linear reader. If you'd rather absorb the information more passively, each design pattern is also downloadable as a poster at anatomyofpublicservices.com/posters.

Chapters 1–3 describe how public services can be designed to work much harder for the public.

'Burden eliminated' (chapter 1) explains the concept of 'administrative burden'. It then describes how proactive, real-time services and the once-only principle for data can lift the burdens of bureaucracy from the public. It argues that a concerted effort to identify and eliminate administrative burdens should become a driving force within the public sector.

'Complexity simplified' (chapter 2) explains how services can use the unique properties of digital to turn complex, disorientating interactions with the state into navigable ones.

'Certainty over time' (chapter 3) describes new types of tools to help people understand their past, present and future interactions with public services. It proposes these as surfaces for automation and prediction to 'scribble upon', without removing agency from the public.

Chapters 4–6 describe the fundamental parts that next-generation public services will be built from: credentials, common components and data.

'Shards of identity' (chapter 4) sets out what is different about digital identity and describes the move from 'I am' to

'I have done' that digital credentials enable. It covers how they can join up services and become the raw materials for more proactive public services and a tool to design against shame.

'Common components' (chapter 5) explains how digital systems are made from components and how they can support multiple public services. It covers how this modular infrastructure changes not only how services get built, but also who gets to build them.

'Data as infrastructure' (chapter 6) describes why the public sector needs to lose its obsession with 'data sharing' (which is akin to photocopying data) and instead manage data as a common resource that supports multiple services.

Chapters 7 and 8 explore the relationship between citizen and state and the implications for the design of public services. Chapters 9 and 10 detail some concrete ways that a digital public sector can understandable, accountable and trusted.

'Empathy augmented' (chapter 7) explains that people don't just care about outcomes, they care about how their interactions with public services make them feel. It advocates that technology should be in service of the relationship between the public and officials, and that the design of services should be anchored in the reality of relationships and objects.

'Designing the seams (not seamless design)' (chapter 8) argues for a recalibration of design practice, so that public services can contribute to a public image of how democracy works and design in co-production as a feature of services, not an upfront activity. Rather than stripping away and simplifying, it describes digital services that have a sense of place, that orientate users and that have routes to reach in and understand the underlying rules.

'Accountable automation' (chapter 9) covers why putting on the public record how digital services and infrastructure work, and how they are changing, will become as critical a part of public life as the publishing of laws and official records. It describes how, because the creation of software is organized around change, accountability mechanisms need to be organized around change too.

'Immunity to treachery' (chapter 10) explains how digital services and infrastructure create new power dynamics, looks at the risks of government being disintermediated by the private sector, and explores what the approaches to services and infrastructure described elsewhere in the book mean for the organization of the work of government.

I have intentionally avoided exhaustive lists of 'best practice'. It is highly unlikely that the exact thing that worked in one place will work in another. So, where they are included, please take them as examples to illustrate the point rather than things to copy wholesale. The other thing about examples is that they go out of date rapidly, so any printed list has a sharp decay curve. That said, I know first hand how valuable examples can be when you need to convince others of an approach (despite all the talk of 'innovation', it's difficult to find people who want to be the first at anything), so at the end of each chapter you'll find a link to a web page with more examples.

Finally, there is a question that I hope you can carry with you through this book: to what extent is the digitization of the public sector inevitable? You may believe that we are in the foothills of the total digitization of the public sphere. Conversely, you may feel there are areas into which technology should just not tread, or that digital services are somehow ancillary to what the public sector does. Hopefully, having this question in mind will help you both calibrate how important you think the ideas described in this book are and think about where they should be applied.

Regardless of where you stand, what I truly hope this book helps you do is build an understanding of the *qualities* and *affordances* of digital public services and infrastructure in a way that you can apply in your work. I hope that it will leave you with a picture of the *possibility space* for public services, and an understanding of where choices lie.

It's time to share the benefits of digital with the public.

CHAPTER 1

Burden eliminated

I n the UK in the 1930s there were riots against 'the means test'. The household means test was an assessment of household income for anyone claiming unemployment benefit, also known as the dole.

Government inspectors could visit people's homes to assess how poor they were. Neighbours were incentivized to report each other to *the means test man*, who might check the state of bedding and kitchen utensils to establish a family's 'means'. One family member taking a part-time job might mean the benefit was refused for a whole family.[1]

The humiliation of the inspections lodged deep in the public psyche, haunting British politics for decades.[2] Britain's system of post-war benefits was, in part, a response to the unpopularity of the means test. The Beveridge report, which set up the framework for the UK's welfare state, proposed a system of social security that was based not on means testing but on principles of insurance and universality.[3] Means testing was, effectively, put beyond use as a policy instrument.*

The means test was an extreme example of 'administrative burden': the cost to people and communities of dealing with bureaucracy. In their book *Administrative Burden: Policymaking*

---

* It was remembered as 'ruthlessly mean' in the 1960s by Geoffrey Howe, who would go on to serve in the Conservative governments of the 1980s, and universality remained a point of principle on the left of politics into the late 1990s.

by Other Means, Pamela Herd and Donald Moynihan define three types of cost that administrative burdens commonly create for the users of public services.

- *Learning costs*: knowing that a service exists in the first place. It's difficult to apply for financial support if you don't know that support exists.

- *Compliance costs*: the effort required to follow the rules. That might include filling out a form, attending an assessment, providing evidence or reporting changes of circumstances.

- *Psychological costs*: how the experience of using a service makes people feel. Do people's interactions feel fair? Do they value the relationship? Is the subject – such as providing evidence of financial hardship or abuse, or reporting information on the death of a loved one – inherently triggering?

The bureaucracies of the paper age created learning and compliance costs for people in understanding and completing paper forms. As those forms worked their way through the mail system and then government back offices, they created uncertainty and psychological costs. As services began to go online, these burdens were frequently replicated form by form; some even retained their old names. In the United States, for example, the W-2 is a paper form issued by employers to their employees, detailing their annual earnings and the amount of tax withheld. In 2024 the United States Social Security Administration still provides an online service for employers to submit digital versions of 'W-2 forms'.[4]

As the title of Herd and Moynihan's book suggests, the creation of administrative burdens can be a policy *choice* – a way to shift the behaviour of the public and distribute outcomes in a way that meets political objectives. The number of successful digital applications for some state-level US health and welfare programmes, for example, is dependent on which political party is in charge in that state, even when the federal government exercises some control over their administration.[5] Unpicking the implicit choices from the explicit ones is a far from simple task.

Some administrative burdens may be choices, but like fish that don't know they are swimming in water, governments are often unaware of the ones they create incidentally. In 2023 it was estimated that 1.3 million UK households were eligible for but did not take up the United Kingdom's main working-age benefit, Universal Credit, with £7.5 billion going unclaimed each year. Nearly 3 million eligible families were not claiming Council Tax support, and 5 million households were missing out on nearly £2 billion of support for water, energy and broadband bills.[6] As of the end of 2023, it was taking up to *two years* for changes to be made to the Land Registry in England and Wales. Despite its digital back-office systems, many of the changes to its database were still instigated by paper form.[7] A 2010 attempt by the UK government to target payments towards those living in fuel poverty potentially missed 40% of eligible people because the people with their names on the bill were not necessarily the people in receipt of benefits.[8] While in 2020, the National Health Service in England ran a bus-stop poster campaign warning of a £100 fine for fraudulently claiming free dental care. It promoted an online questionnaire for people to manually verify if the exact mix of government benefits they were receiving entitled them to free dental care or not. But this was information the UK government already held![9]

Better use of technology and data could have all but eliminated these issues. But instead, administrative burdens continue to have real effects: fragility in the housing market, families with less money, holes in teeth, cold children. Sadly, these are far from isolated incidents. Despite decades of digitization of public services around the world, governments have not been sharing the benefits of digitization with the public. In the rush to digitize public services, there is little space to ask: *why bother?* If there is an answer, it is often financial: savings from automation, less duplication, the replacement of neglected systems or the closure of call centres. The aim of most digitization programmes is the status quo, delivered more cheaply. This is not surprising. Government business cases are woven from such hopes. The resulting documents are catnip to treasury officials. But efficiency is a trap.

Even where services are well designed, it is all too easy to make things simpler and cheaper for government while services remain fundamentally the same for the public (or are even made worse!). That's because the things that are expensive or complex for government are not necessarily the things that are expensive and complex for users. Government officials have strong incentives to fund work that makes a service simpler for users, but only when the existing process is expensive. Services that reduce burdens for users but cost government money, such as digitizing appeals processes or automating enrolment, are overlooked, as are burdens that exist *between* services, as the bus stop scare campaign illustrates. The result is that there are many problems that digital is never applied to.

Rather than seeing digital as a short-term cost-saving measure, the aspiration should be to share the benefits of automation with the public. Applied correctly, digital can reduce the burden on individuals, families and communities to close to zero in everything from buying a house, to claiming benefits, to accessing free childcare. The opportunities to do this are better today than they have ever been.

In 2011, when we were designing the first version of GOV. UK at the UK Government Digital Service under the project name 'alphagov', we had a set of design principles pinned to a column in the middle of the office. Some of those principles stand up pretty well today: things like 'Set clear expectations' and the reminder to design for context by being 'Consistent, not uniform'. Others seem very dated, particularly 'Google is the homepage' and 'Every visit is a new user'.[10] Those two principles made sense back in 2011 because

- web search was becoming the primary discovery mechanism for public services;

- the back-office processes were generally digitized versions of paper processes; and

- services did not have access to data about a user's previous interactions – services did not 'maintain state', as it is termed in computer science.

Interactions with public services were *active* and *transactional*. Users had to *actively* identify what they needed to do, and then complete a form, be it on paper or digitally. The transaction was 'done' when a licence was issued, an account updated or a letter posted. So if you wanted to make things better for users in 2011 (as we wanted to with GOV.UK), more clearly written web pages, a consistent design system and the removal of superfluous questions from digital forms was a good place to start.

This was not unique to the public sector, either. Many commercial services were similarly optimized for web search and transactional interactions. When I worked at the digital printing start-up moo.com in the mid 2000s, we similarly designed for a world where every visit was a new user. We designed 'tools for users without experience' and launched with no way for users to sign in and view their previous orders.[11]

Things are different now. Rather than being active and transactional, we are moving towards a future in which people's interactions tend more towards *passive* and *real time*.

Passive interactions are already common in commercial services: spam filters in email; recommendations in streaming services; deliveries from online retailers being bundled together;

a web browser that gives automatic suggestions for filling in contact details. Increasingly, it is the service that does the work, *proactively* anticipating a user's needs.

Real-time interactions are also common. Cars are shown moving on a map in ride-sharing apps, and prices change based on demand. If you want to buy something online, you can see immediately if it's out of stock. Making payments using a mobile wallet triggers an instant notification that the transaction has taken place.

Real-time, passive interactions are beginning to appear in public services too. More than 8 million Ukrainians have used the eAid service to access social security payments without having to fill in extensive forms. The service's eligibility checks are done automatically, using data from across government.[12] In India, when a user updates the address on their Aadhaar identity credential, they are prompted to copy the change to their digital driving licence and to other credentials too.[13] In Estonia, digital infrastructure for real-time financial reporting and invoicing between companies is being developed under the banner of 'the real-time economy'. It aims to do away with the need for paper receipts and invoices when companies interact with each other or with the government.[14] Estonia also operates a 'once-only principle', so that data entered in one service does not have to be entered again in a second service. In fact, services are forbidden from requesting data that is already held by government (see chapter 6).

These examples show the potential of digital and data to remove administrative burdens. However, just because administrative burdens *can* be identified and eliminated, that doesn't remove the fact that they remain as a *policy choice*. In fact, digital probably makes that policy choice easier and cheaper. New compliance costs can be targeted at users based on the data held about them; new demands can be made, and made more often, using digital channels than is possible with analogue ones. Just as mobile devices and digital communications mean it can feel like you are always at work, digital services could make it feel like the government is always setting you jobs – like you are always being governed.

The United Kingdom's digital welfare service, Universal Credit, shows how digital services can simultaneously create and remove burdens. It also illustrates how the funding of digital work in the public sector – and how the application of digital-age design to public services – can prevent the benefits of digital being shared with the public.

Universal Credit is a single working-age benefit for people who are either unemployed or are in work but in need of financial assistance. Like the means test of the 1930s, it is a household benefit. From its inception, it was designed to be complex: the aim was to allow policymakers to reward or punish detailed combinations of circumstances and behaviours. Despite its name, Universal Credit is at the other end of the policy spectrum from the universal benefits of Britain's post-war years. It is a *hyper*-means-tested benefit – one that is only really practical when mediated via a digital account. It requires many data points to maintain a claim, and it uses that data to calculate a bespoke payment amount and activity regime for each claimant.[15]

Universal Credit was controversial as a 'digital-first' service,* but most of the burdens it created are a result of policy choices, not technology. The policy, which attempted to 'mirror a world of work', assumed that people were paid monthly and would therefore have a month's worth of wages in the bank if they became unemployed. It assumed that couples exercised shared control over their finances, so it was a single household payment. It assumed decisions about the number of children in a household were the product of financial planning, so parents with more than two children were not to be given additional support. Furthermore, because it assumed that families could save money by

---

* From the start, digital channels were intended to be the main way for the public to claim Universal Credit.

moving to lower-cost accommodation if they had more bedrooms than children (something that is seldom true in the United Kingdom), it created a financial penalty for 'spare rooms'. Government caseworkers also set users tasks, such as searching for jobs or updating their CVs,[16] with the assumption being that these activities would result in better outcomes for users.

A charitable interpretation would be that these measures were based on naive understandings about the reality of normal people's lives: an idealized understanding of employment, family finances and housing. Less charitably, they were an attempt to form society to a particular model. Either way, the result was a system that put the majority of users in debt to the government before a single payment had been made because they had to apply for an additional advance payment. Where people's earnings did not neatly match the monthly ideal, the payments they received became lumpy and unpredictable.[17] Users were also expected to understand the need to report more than fifty different changes in their circumstances, including things such as a change of address, starting cancer treatment, a child becoming disabled, or starting full-time education.[18]

Long after the policy had been written, I was part of a small, multidisciplinary Government Digital Service and Department for Work and Pensions team tasked with creating a workable implementation of the Universal Credit policy after an initial attempt had floundered.[19] Nearly £700 million had been spent attempting to get the policy up and running, using outdated outsourcing approaches, with little to show beyond a £196 million bill for written-off 'IT assets'.[20] My job for just under a year was to attempt to create a coherent service design framework for Universal Credit.[21]

This was the summer of 2013. GOV.UK had gone from being an 'alpha' proof of concept to becoming the single place in which to find government services and information from all central government departments. That spring it had won the Design

Museum's prestigious Design of the Year Award* for making the public's interactions with government simpler, clearer and faster.[22] However, the approach we had used for the design of GOV.UK proved an incomplete fit for Universal Credit. To explain why, it's necessary to take a brief historical detour to the origins of user-centred design.

GOV.UK had been designed around the concept of 'user needs'. User needs are a core part of user-centred design practice: an approach that prioritizes iterative improvement based on an emergent understanding of how users interact with a service. They are the things that a service must satisfy for a user to get the correct outcome. To get a visa, a user *needs* to understand what documents to provide. To avoid a fine, a user *needs* to know the tax return deadline. 'Start with user needs (not government needs)' was number one on the Government Design Principles list.[23] Many a laptop was adorned with a sticker asking: 'WHAT'S THE USER NEED?' Government ministers even talked of 'user needs' in parliament.[24] It seemed radical but, by the 2010s, these were already old ideas.

The British social scientist Enid Mumford started her career in industrial relations research in the 1950s. She spent months underground talking to miners at the coal face to understand how new technology affected them, and she took jobs in canteens at Liverpool docks to understand the working practices of stevedores. She would go on to apply the same approach to computer systems, conducting research with the users of newly computerized organizations, such as banks and government departments.[25] In 1975 Mumford developed the 'needs-fit' theory of job satisfaction. This said that computer systems should be built around the working practices and preferences of the people using them. Successful system design, she argued, required a fit between the needs of the users and the capabilities

---

* It beat London's 'The Shard' skyscraper to the title and was dubbed 'boring.com' by one newspaper for its pared-back design.

of the technology. Failure to involve the users of systems meant that projects had a high risk of failure.[26]

A couple of years later, in 1977, the idea of 'user-centred design' was first formulated by the American computer scientist Rob Kling, citing Mumford's work. Programmers would often, he said, 'impose systems' on users.[27] This occurred when they were 'explicitly insulated from close contact with computer users' and when they viewed themselves as 'change agents "reforming" an inefficient organization'.* Having user needs as the organizing design principle for GOV.UK had worked because the proposition was inherently task based: find the right content and services as quickly as possible. The design response was also clear: remove superfluous pages, make the 'calls to action' as clear as possible, and have a consistent design so users don't have to relearn how things work each time.

However, Kling had also noted that user-centred design assumed that users don't have 'incompatible interests'. This misalignment of needs is not uncommon in public services, and Universal Credit, as a policy, has many of them. For example, Universal Credit's need to categorize the relationship status of two partners in a household to calculate their benefits might be incompatible with that couple's need to negotiate the formality of their romantic relationship, the division of caring responsibilities for their children, or the management of their finances. For Mumford too, needs had been part of a wider approach to participation, which she described as 'democratic processes that enable employees to exercise control over their own work environments and work futures'. These included things like software design groups that were *elected* by the staff who had to

---

* This described the UK government prior to 2012. The teams creating digital services were insulated from contact with users (as were ministers and civil servants). News stories about poorly designed systems and large overspends on technology projects were common. Government ministers would regularly launch new services on the morning news only for them to crash twenty minutes later.

use a system.[28] But Universal Credit didn't have a mechanism for users to negotiate what burdens were appropriate.*

Attempting to design for a system that intentionally created burdens for the public using user needs didn't really work – at least not on its own. Rather than user needs, the Universal Credit digital account was created around the management of the fluctuating administrative burdens the service placed on users. It was organized around a 'to-do list' and a 'journal'. The to-do list aimed to make sure that compliance costs were as clear as they could be to users, and it placed an obligation on the Universal Credit service to notify users when they needed to do something, rather than the other way around. The journal provided a clear record of a user's interactions, and because users were also able to add things to their journal, they could tell the service something on their own terms, without having to internalize every rule.

Universal Credit may have created many new burdens, but it also showed the potential for automation to remove them. Earnings information was sourced automatically from income tax records, reported by employers. This meant that, so long as a user's work pattern aligned with the expectations of the policy, there were no compliance costs for users in reporting their earnings. Despite the many documented problems with the policy, the concept of using earnings data to calculate benefits automatically is one insight from the architects of the Universal Credit policy that will probably stand the test of time. Their assumptions about work and families are destined to be remembered with the same fondness as the 1930s means test.

However, in line with the business case for Universal Credit, automation has mostly been used to reduce the cost of administering the welfare system.[29]

---

* Outside of elections anyway.

More automation in Universal Credit could automatically suggest the best payment cycle for a user based on their previous income. It could pre-fill forms to apply for related benefits and support. The reporting of changes of circumstances could become a more passive task: for example, users could be prompted to confirm that a child's education status has changed at sixteen – a moment when that is likely to be the case. The system could accept digital credentials from childcare providers and banks that remove the need for data to be reported manually.

The case for 'automation for the public good' described above for Universal Credit is mirrored in countless public services around the globe.

We have the opportunity to move from a world where, rather than users searching for things, there can be an expectation of things coming to users. 'Proactive services' could take action on users' behalf, surfacing recommendations and automating renewals. Where appropriate, interactions could be initiated automatically or with minimal involvement from a user. Reusing and exchanging data between services could mean fewer forms, and real-time services could mean decisions can be made more quickly and brought closer to the point of use. Historically, this would have been described as 'removing red tape', but red tape is not a problem if the rules are handled by machines! Every time a form that could be filled in by a machine is not, we should see that as a theft of time from the public, and we should regard pushing the task of understanding eligibility onto the public as blocking people from accessing the support they require.

The opportunity offered by the next generation of public services is nothing less than the systematic elimination of administrative burden across society – the automation of the mundane, everywhere.

PATTERNS

1. The once-only principle
2. Proactive services
3. Real-time services

The once-only principle

The once-only principle makes data entered in one service available for use in other services. It saves users time by reducing the number and the length of forms that need to be completed. So, if someone moves house, changes their name, starts a job or is granted citizenship, they don't have to report the same information again and again and again.

The diagram shows the once-only principle in action for someone who has moved to a new country for work. Having completed a visa application before they left their home country, they registered for social security shortly after arrival so that their employer could add them to the company's payroll system.

A few weeks later, they need to register for income tax. Their employer's details, new address and visa information are all available to be used by the tax service. Once they have confirmed the information is still correct, they only need to add the information unique to their income tax registration.

There are two approaches to implementing this principle.[30] In the first, data is held by multiple government agencies. When a change is made in one place, it cascades to other agencies. For this approach to work, agencies need a way to subscribe to changes in each other's data. For example, a tax agency requires a way of signalling it is interested in data about employers. In the second approach, data is stored once and accessed directly by multiple government services.[31] For this approach to work, data needs to be managed as a common resource. (See the 'Data access, not data sharing' pattern in chapter 6.) Regardless of approach, the experience for a user should be the same: information shouldn't need to be entered twice.

The once-only principle is just that: a *principle*, not a service in its own right. It is something designed *into* services. So, if someone moves home, rather than using a specific 'Report a change of address' service, they are given the option of updating their address from whatever service they tell first.

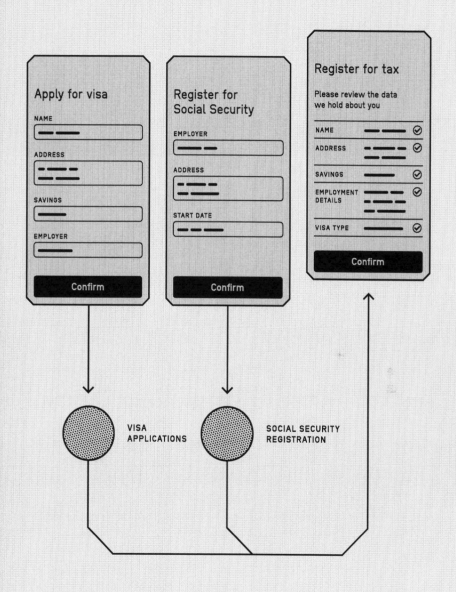

That said, safeguarding concerns, data protection and privacy must come before convenience. Services need to explain what is happening; make clear the providence of data; and, where appropriate, seek explicit consent. When users are worried, they should have clear opportunities to seek help or opt out. Failure to prioritize these things risks creating emotional costs. A user may find themselves asking, for example: 'If I update my new address before I move country, what will it do to my work visa application?'

Applying the once-only principle systematically means changing how data is managed across the whole public sector. Even then, implementing the once-only principle may not be a priority for every organization responsible for operating a public service. One option is to make the once-only principle a public right, although this should only be considered once the approach has been tried and tested.

Proactive services

All services are built on rules. Even universal services have some sort of eligibility requirements (a five-year-old doesn't get a state pension). Proactive services anticipate users' needs by combining data with eligibility rules. Users spend less time trying to understand their entitlements and less time filling in forms.

The diagram shows three types of proactive notifications that the owner of a residential care business might see. The first notification shows that the company's care licence and the owner's driving licence are due to be automatically renewed soon. The second shows that the company will probably need to register for sales tax. The final one indicates that the business has qualified for local tax relief because the property has been revalued.

For most rules, the opportunity for *full* automation is limited. Eligibility requirements often have exceptions and caveats that are best represented in words, not code. The data held may not line up 100% with the eligibility criteria for other services,

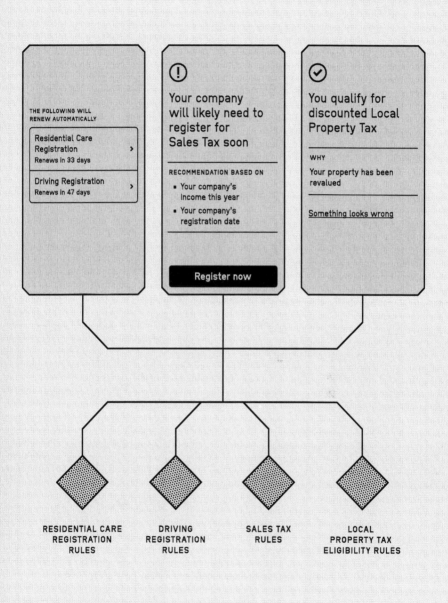

so extra information might need to be collected.[32] Different agencies may also use different definitions for things such as 'disability', 'a household', and what counts as 'income'. Rather than a bug, this messiness can be a feature. Policymakers need to be able to direct support to particular groups and contexts.

However, even where rules cannot be fully automated, proactive services can make useful *recommendations*.

Government websites tend to use language like *may* and *might* to describe eligibility. That's about as committal as most have been able to get historically. But because proactive services have access to more data, they can offer a higher degree of commitment, telling a user if they will 'likely' or 'probably' be eligible for something.

For example, it may be 'highly likely' that a parent of a disabled child who already claims welfare, with an income below a certain threshold, is eligible for additional support as a carer. A proactive service could notify them and even pre-fill parts of the application form. Data-driven nudges like this could help more people access the services they are entitled to and get the support they need.[33]

'Likely' and 'probably' create a safer space for AI to operate in than full automation. AI has a tendency to make stuff up and give different answers each time for the same inputs. Applying AI to high-stakes domains like tax and welfare, where the law is fixed and deterministic, is therefore probably undesirable. But what if AI's task was not to determine but just to recommend? AI with knowledge of every government policy document might be able to surface valuable recommendations to users. (See chapter 3 for how this might be applied to a cross-government task list.)

Real-time services

Real-time services help users understand their eligibility, responsibilities and rights more quickly. They reduce the amount of form filling and can speed up decisions.

Transactional services (legacy)

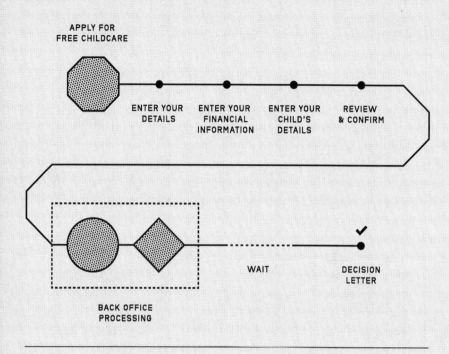

APPLY FOR FREE CHILDCARE

ENTER YOUR DETAILS

ENTER YOUR FINANCIAL INFORMATION

ENTER YOUR CHILD'S DETAILS

REVIEW & CONFIRM

BACK OFFICE PROCESSING

WAIT

DECISION LETTER

Real-time services

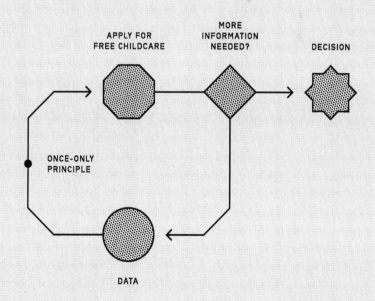

APPLY FOR FREE CHILDCARE

MORE INFORMATION NEEDED?

DECISION

ONCE-ONLY PRINCIPLE

DATA

The diagram contrasts a legacy 'transactional' service for applying for free childcare with a real-time one.

Under the transactional service, the user fills out a digital application form. Once submitted, it is processed by back-office staff who check if the application is valid and then decide if the child is eligible for free childcare. After a potentially lengthy wait, a letter notifies the user of the decision.

The real-time service combines data that is already held with automated eligibility rules to make a decision. If all the data needed to reach a decision is available, then a decision can be made immediately. If not, only the missing information is requested. Where a manual check is required, e.g. if back-office staff need to check a paper birth certificate, then there is nothing preventing a user from progressing with other tasks. The progress of manual checks can also be shown in real time (unless it is a fraud or security check). Once complete, the service issues a digital credential (rather than a letter) that can be shared with childcare providers straight away.

Real-time services are a fundamentally different model from transactional ones. With transactional services, nothing can happen until a validation application has been submitted. In a way, the service assumes a user is in the wrong until they have done the work to prove otherwise. Real-time services put the onus on the service to do the hard work.

STRATEGY

Share the benefits of digitization with the public

Today's restriction of automation to areas where business cases claim that proximate savings are to be found leads to poor outcomes for the public. The aim of automation should not just be to make public services more efficient, it should be to systematically remove administrative burden from across society. This requires changes to how digital work is funded, how services are designed, and what users should expect from public services.

Because administrative burden can be a policy choice, it has to be a policy choice to identify and remove it. Public sector leaders must make it a clear policy intent to systematically identify and remove both intentional and unintentional administrative burdens. This will require building both a common understanding of where administrative burden lies and a consensus on how to measure it. The biggest opportunities lie in simplifying the complexity *between* different parts of government (see chapter 2), so it is not enough to identify administrative burdens within individual services, or indeed within the bounds of one government agency or another. It must be a system-wide effort.

Where administrative burden remains a policy choice – as it does when unemployed people are required to prove they have been applying for jobs, for example – it should be clear that the decision has been made and it should be a matter of public record how it will be tested.

The elimination of administrative burden should also become the primary concern of those designing public services. 'User needs' is a critical concept in understanding how to make services work for people, but it is too limited on its own. The experience of Universal Credit shows that designers of public services must not only ask 'What is the user need?' but also 'Where are the administrative burdens?', even where they exist outside the remit of their service. They should also measure the number of interactions they are automating, in comparison with older non-automated interactions (such as apply, cancel and renew). There needs to be design effort too in creating the new interaction patterns needed to support automation (some of these are explored in other chapters).

But design must also go hand in hand with an unapologetic use of technology to pursue automation for the public good. It is a common refrain that digital transformation is 'not about the technology', but administrative burden is one area where that is definitely *not* true. Designing the best possible services means considering the capabilities of technology. Implementing the approaches described here is first and foremost a technology intervention, not a design one. You can't design services that reuse data or make real-time decisions in any systematic way if you don't have the infrastructure to support doing so. These changes require data to be treated as a *common* resource and accessed over the sort of *common* infrastructure that countries like Ukraine, Estonia and India have implemented.

Eventually, we can imagine these approaches crystallizing into a general 'right to prediction' and a 'right to real time', alongside a 'right to tell government once' (which has been implemented, to some degree, in Estonia and a few other jurisdictions). But any such mandates must always follow the practical – telling must follow what works.

Main points

1. Administrative burdens create learning costs, compliance costs and emotional costs for users.

2. Digital services have often replaced burdens one for one.

3. Administrative burdens can mean policies don't work as intended.

4. Administrative burdens can be a policy choice.

5. The aim of most digitization programmes is the status quo, delivered more cheaply, but efficiency is a trap.

6. The next generation of services are real time and passive, rather than delayed and active.

7. The once-only principle can reduce the number and length of digital forms people have to complete.

8. Proactive services can automate or semi-automate eligibility checking and renewals.

9. Real-time services can reduce waiting and uncertainty for users.

10. The systematic elimination of administrative burden should be the driver for digital public services.

Examples: anatomyofpublicservices.com/examples/burden

CHAPTER 2

Complexity simplified

Services too often reflect the organigram of the organization that made them: banks with separate phone lines for each department, a multitude of websites and apps for different government agencies, chasing test results lost between different hospitals. Bureaucracies *are* complex, but people should never *have to* understand their structure to get things done. This does not mean that structure should be totally hidden (see chapter 8), but understanding it should never be a precondition of the use of a service.

This phenomenon is neatly summed up by Conway's law, which was proposed by the US computer scientist Melvin Conway:

> Any organization that designs a system (defined broadly) will produce a design whose structure is a copy of the organization's communication structure.[1]

Conway's law is particularly pronounced at the scale of a nation state. The shape and composition of each country's public sector are the products of constitutional arrangements, happenstance and evolution, but they are nearly always complex organisms made from many parts.

The number of core departments and ministries is generally not very large: the German federal government is divided into

fifteen ministries; the United Kingdom has twenty-four ministerial departments and twenty non-ministerial ones; South Africa
has thirty-nine national departments. The list of agencies tends
to be longer: Germany has nearly 300 federal agencies of various types; the United Kingdom has more than 420 agencies and
'public bodies'; South Africa has nearly 200 'public entities'.[2]

Add in local and regional government, courts, police, public
utilities and others and you get a very high level of structural
complexity (the number of organizations and the relationships
between them) and functional complexity (the variety of services the public sector provides). Layer on the separation of
powers and constitutional checks and balances and it is more
complex still. As Conway's law predicts, public services tend
to mirror these structures. Complexity creates voids between
services that are filled by, for example, accountants, brokers
and lawyers, often at a cost to the end user. Whole markets are
created and sustained in this way.

Complexity also creates a space that can be manipulated
and exploited by scammers and rent-seekers. It creates costs
for government agencies when people's mental model of who
provides a service doesn't align with reality.* People's lives don't
reflect the organizational complexity of government, so the services they rely on shouldn't either. Services should be designed
with the grain of people's lives, not against it.

Despite the computer science origins of Conway's law, digital
services offer the best opportunity to do just this. Digital services can collapse the distinctions between organizations in a
way that is unparalleled at scale and cost when compared with
traditional professional services.

Think for a minute about ride-sharing services such as Bolt,
Grab or Uber. They connect passengers with drivers near-
instantly and regardless of their location. Services such as

---

* An example would be people assuming they should contact the tax agency about a
bereavement payment when they should in fact contact the benefits agency.

Skyscanner let users find and book flight tickets with multiple airlines. Similarly, Amazon allows customers to purchase almost anything, from millions of sellers, with more than 2,000 new vendors added every day, each a separate business.[3] The different warehouses, sellers and fulfilment providers all still exist in the world, but at the point of use, any distinction is abstracted away.

Digital services change our relationship with space and time. Among geographers, this is referred to as 'time–space compression', but it is more widely familiar in the saying 'technology makes the world smaller'. Time–space compression appears to make the world smaller because technology allows for faster communication between distant places.[4] You don't need to visit a taxi office if you can book a taxi online, for example.

There's a related idea, 'time–space distanciation', that describes the 'stretching' of relationships and networks over time and space. Technology allows us to interact with a widely spread network as if nodes of that network were co-located.[5] You don't need to phone three different taxi companies if you can interact with all of their drivers in a single app.

In digital services, the complexities of time, location and organizational boundaries become materials to be designed with – things that can be dialled up or dialled down in service of the needs of users.

This is what we did when we created GOV.UK: a single website for publishing information and finding services. When it launched as a beta in early 2012, it replaced one government website, Directgov. Later that year it replaced a business-facing website called Business Link, and later still six government departments moved their websites to GOV.UK. By April 2013 all twenty-four central government departments had moved over.[6]

GOV.UK rapidly became one of the most visited websites in the United Kingdom and set new standards for accessibility and design. Each organization that joined GOV.UK still published information, but now there was a single place for the public to search for that information and find services. By 2015 it had replaced 1,882 organizational websites, including those of government agencies and embassies.[7]

Conway's law is felt acutely in the midst of something life changing, like when a child is born, a business is started, someone leaves the care system, or a company wants to start exporting. In these situations there is no escaping interacting with multiple parts of the government organism.

GOV.UK may have put government publishing in one place, but services that cut across organizational boundaries tend to be the exception. However, that has begun to change.

In 2019 Estonia began to design new services around 'life events', such as having a child, that transcend the boundaries of government agencies.[8] Singapore's GovTech agency is following a similar approach and has developed the LifeSG 'Register your child's birth' service.[9] This helps new parents navigate multiple government services, including applying for a baby bonus payment and applying for a child's library card. The Italian government has recently begun trials of what it calls 'mobility as a service', the long-term aim being a single nationwide app for finding and paying for public and private transport. Ticket prices and routes will still, presumably, be set by individual providers, but the app will make it feel like a single service.[10] These examples all *abstract* away the distinctions between organizations.

In addition to abstraction, a second way in which digital can mitigate the effects of Conway's law is by breaking the one-to-one relationship between the point of delivery of a service and the underlying rules. APIs expose the rules of a system in a way that can be used in other software programs.

The idea of APIs has roots in an early computer: the British EDSAC of the late 1940s. Reusable subroutines that solved common problems, such as calculating a logarithm, were stored on punched paper tape and organized in a filing cabinet along with instructions on how to use them. The complexity of writing software was reduced because the common code had already been written and could be used in multiple programs. In 1951 computer scientists Maurice Wilkes and David Wheeler, who worked on the EDSAC's modular software library, documented the subroutines in what was the world's first ever programming book.[11]

As programming languages evolved, it became standard to have well-documented APIs for interacting with the underlying computer. But rather than physical filing cabinets of paper tape, the APIs were integrated directly into the language. In the mid 2000s a similar pattern became common in so-called Web 2.0 services. Those services provided APIs that third-party software could use to extend the functionality of the service. eBay and Salesforce had the first significant web APIs, but the gold standard was the photo-sharing website Flickr.[12]

The Flickr API allowed third-party software, written by software engineers that the Flickr team had probably never met, to upload photos, search, post a comment or curate a gallery. Pretty much everything a user could do on the Flickr website was replicated in the API. It was easy for software engineers to use, had good documentation, a clear process for managing changes, and example software code in the popular programming languages of the day.

Some services built using the Flickr API replicated the functionality of Flickr itself, e.g. letting people upload and browse photos from their Nokia phone or Windows laptop.[13] But many were entirely new: services that printed business cards or created comic strips from photos of kittens, for example.[14]

Blogging in 2006, Flickr cofounder Caterina Fake explained how the Flickr API meant that they could build partnerships with start-ups who wanted to build on top of Flickr that they would otherwise not have had the internal capacity to engage with.[15] Its API helped Flickr to grow and to meet needs that otherwise would have gone unmet.

Rules APIs break the relationship between the organization responsible for the 'brains' of a system and the service a user interacts with. They are a way of meeting more user needs, more of the time. It is now common for digital services to expose their internal rules in this way. Users of the running app Strava can upload their runs from different brands of smartwatch, for example, and users of digital banks such as Monzo can compare spending from their bank account and third-party credit card providers in a single app.*

In the public sector this has a potentially radical effect. Typically, policymakers have had to choose which layer of government (central, regional or local) should operate a service. They have also had to make a choice between public, private and third-sector delivery. A single public-facing service had to try to meet the needs of all users. APIs turn these into false choices because multiple public-facing services can use the same underlying rules, even if central and local government might offer different versions of the same service, e.g. a user could choose to apply for welfare payments on a central government website or a regional government app and get the same outcome.

Rather than having to decide between privatizing or outsourcing a given service, governments can follow both a 'wholesale' and a 'retail' approach to meeting the needs of users.[16] Governments should generally still provide a 'retail' version (a government-operated service provided directly to users), but rules APIs mean that commercial and third-sector organizations can provide services too (the 'wholesale' versions). Complementary, overlapping services can be provided by any layer of government and by commercial or third-sector organizations. It's OK when they overlap, complement and duplicate each other.

Recreation.gov provides a nice example of this. Operated by the US federal government on behalf of multiple agencies, Recreation.gov is a service for visitors to federal land that helps users find and book campsites and tours. However, because it is built on an API, users can also book via commercial vacation-planning services too.[17] There are multiple ways to get the same outcomes because of the common rules provided by the API. Similarly, the

---

* This is enabled by Open Banking, an industry-wide API standard.

United Kingdom has APIs for retrieving food safety ratings for restaurants that are used by delivery apps such as Just Eat.[18] Alternatively, people can look up the same information on the Food Standards Agency website.[19] These government and commercial services overlap and, to some extent, compete, but it doesn't matter if the aim is to maximize the number of people accessing public land or to minimize the number of people getting food poisoning.

The Making Tax Digital initiative by the UK's HM Revenue & Customs is an example of a wholesale-only approach. Its APIs allow software applications to interact directly with government systems, enabling businesses and their accountants to report financial information in close to real time. Rather than forcing companies of wildly different sizes and types to use the same piece of software, tasks such as submitting VAT returns or income tax updates can be done directly from whatever bookkeeping software the company is using. At the beginning of 2024 there were 559 different integrations listed on the UK government website, each serving a slightly different audience.[20]

The APIs developed by the United States Department of Veterans Affairs provides a good example of third-sector use of APIs. Registered veterans-support organizations create services that use the APIs to help their members apply for benefits.[21] Services like these promise a future in which users get the support they need from charities in new ways. Users will be able to securely give citizen support charities access to things such as benefit claims, immigration status and school applications without, as often happens currently, carrying a life's worth of papers around with them.

There will always be needs that government cannot meet (because of capacity or access to users) or will not meet (because of political constraints). Rules APIs are a way of meeting those needs.

However, there are limits to how much the design of services can simplify the complexity of the public sector. The reality of people's lives is that they are messy, and the designers of services can never know the shape of that messiness up front. Not every service can or should be designed from one end to the other. Doing so would assume that the design process can anticipate the needs of every user in every situation.

During the Covid-19 pandemic we saw a different type of service emerge: the composite service. Teams across the UK public sector rapidly created digital services for things such as PCR testing, booking vaccinations, ordering home testing kits, exposure notifications and authenticating with healthcare providers. Partly because these were hastily put together, and partly because of local variation and rapidly changing circumstances, these services were often made from different parts, loosely joined together to support different user journeys.

Getting a PCR test, for example, included

- booking a test on a local government website,
- visiting a test site operated by the local authority and presenting a QR code,
- scanning the barcode on an NHS-branded test card,
- recording the details of the test on GOV.UK,
- authenticating using an NHS login,
- using a central government branded physical test kit,
- scanning another QR code to register the test, and
- getting notified of the test result via the central government GOV.UK Notify component.

There was no single end-to-end service here. Instead, there were multiple paths through a series of components and services that were designed to be porous and linked together. Broadly, this is a service design implementation of the robustness principle: a design guideline for software that says 'be conservative in what you do, be liberal in what you accept from others'.[22]

There are other examples too. Ukraine's eBaby service for new parents is a combination of nine services, glued together by digital identity and data infrastructure.[23] And in India users frequently 'jump out' of a service to authenticate with the Aadhaar identity system or retrieve a digital credential from the government's DigiLocker app.

There are also parallels here with how Apple appears to be developing its operating systems. With iOS 13 Apple introduced the Shortcuts app, which provides a visual programming interface for joining together parts of apps into workflows. For example, a 'Leave for home' shortcut might display the weather conditions from a weather app, list some items from a shopping list app to pick up on the way home, and display the quickest bus route for getting home. iOS 14 then introduced the idea of AppClips: lightweight versions of apps with minimal functionality, accessed by scanning QR codes and NFC tags, such as when unlocking a hire bike or joining a cafe's loyalty scheme. Both of these features had the effect of unbundling third-party apps into their component parts.

In 2024 generative AI can already accurately give you the steps needed to create complex Apple Shortcuts. (Try typing 'tell me how to create an Apple Shortcut for leaving the house' into an app such as ChatGPT or Google Gemini.) It wouldn't be much of a jump for generative AI to also start creating user interfaces.

Today, if you type 'What are all the possible government services I might register a newborn child for in the UK?' into a generative AI, you get back a convincing and fairly complete list of steps. What might happen if it could go one step further and assemble a composite service too? Governments that successfully unbundle their services into components and rules will be able to create bespoke, composite services that meet every user need and make Conway's law a thing of the past.

PATTERNS

1. Abstraction services
2. Multiple, overlapping services
3. Composite services

Abstraction services

Digital services can help users interact with multiple organizations at once. Rather than having to maintain a separate bureaucratic relationship with each organization, a single service can manage all those relationships for the user. We call these services 'abstraction services'.

The diagram shows a service for starting a business. It uses abstraction to help users create a company that can start *doing* business, rather than just creating a legal entity.

First, it registers the legal entity at the company registration office. The ownership and naming rules verify that the company name is valid and that the user is eligible to be a company director. The new company is then recorded in the register of companies, and a company registration credential is issued immediately.

Using the new company credential, the service can register for sales and company taxes. The tax office's rules check the registration requirements for the type of company, then update the sales tax and company tax databases. A credential is issued for both types of registration.

Finally, the service applies for licences from the city government. For example, if the company were a cafe, it might apply for a pavement trading licence.

In this example, the mandates of each organization are respected since they maintain the data, set the rules and issue the credentials. The tax office still maintains the definitive lists of who is registered for different taxes. The city government still decides who can have a licence, and it can revoke them when the rules are broken. The complexity of joining everything together is handled by the service.

This type of service is impossible or unreasonably expensive with analogue processes. A paper form can only be in one place at a time, and paying for professional services such as accountants and lawyers to help 'glue together' disparate services is expensive.

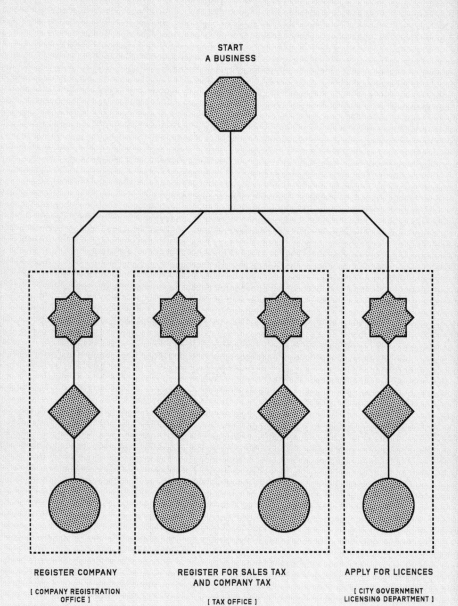

**START
A BUSINESS**

REGISTER COMPANY

[COMPANY REGISTRATION
OFFICE]

**REGISTER FOR SALES TAX
AND COMPANY TAX**

[TAX OFFICE]

APPLY FOR LICENCES

[CITY GOVERNMENT
LICENSING DEPARTMENT]

The diagram is simplified for clarity, but the service might also handle things such as registering as an employer or for payroll taxes, paying local property taxes, or registering for a food safety inspection.

If, unexpectedly, the cafe becomes a social media hit and starts selling cookbooks and merchandise overseas, it would need to register for sales tax with foreign governments too.[24] The service could register for those taxes too. Services that help companies deal with the complexity of rules not just in their home country but also abroad are an interesting growth intervention for policymakers.

The ability to abstract away the distinctions between organizations means services can be designed around the needs of users. In addition to starting a business, events such as having a child, moving house, becoming unemployed, leaving prison or managing the affairs of someone who has died all require people to interact with multiple organizations, often at a time of stress.

'Abstraction services' should be reserved for life events that are common in society, that have clear boundaries and for which users' needs are fairly consistent. Trying to design an abstraction service where user needs are inconsistent, uncommon and have messy boundaries risks creating an inflexible system – or, as often happens, the design process becomes so large and unwieldy that the service is never built because the team designing it doesn't know where to draw the line. In fact, the number of situations it suits may actually be quite small.

Multiple, overlapping services

Multiple services can make use of a common set of rules. Services can be provided by any layer of government, by private companies or by charities, and it's OK when those services overlap, complement and duplicate each other.

The diagram shows four different services built on top of eligibility rules for different types of cost-of-living support. There

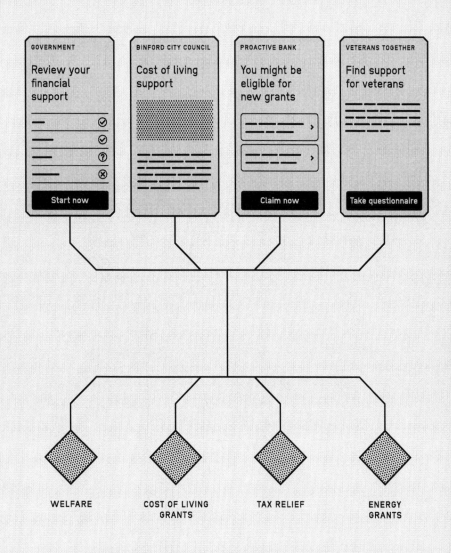

is a central government service, a local government service, a banking app and a questionnaire from a veterans' charity.

- The central government service uses the data it holds to tell users about welfare payments, tax relief and energy grants that they will likely be eligible for.

- The city government service does something similar, but it signposts users to local community-based support as well. It also recommends support that does not align with central government policy.

- The banking app suggests support based on users' spending, without revealing their individual transactions to the government.

- Finally, the veterans' charity uses the rules in a questionnaire as part of a campaign to ensure that ex-service personnel are claiming their entitlements.

Rather than competing, these services complement each other. They ensure that the needs of every user are met, and met in the best place. The best place to do something will not always be a government service. For example, in addition to being able to buy a parking permit from a local government website or office, it might be more convenient for some users to buy it from within the checkout screen of a holiday rental app, at a car rental garage or in a car's maps app.

A city government providing its own version of a service allows for local knowledge and accountability. Local services that do not align with central government policy are not uncommon. In the United Kingdom, central government has avoided referring people to food banks, while North American 'sanctuary cities' provide access to services for migrants without cooperation with federal immigration authorities.

Breaking the one-to-one relationship between a service and the underlying rules means that every user need can be met, and met in the way that is most convenient for a user. This type of variety of service provision is possible because there is a single source of truth about the underlying rules.

Composite services

Composite services are not *designed* to solve *whole* problems for users: they are made up from loosely connected parts that combine to meet the needs of users. Those parts might include elements of other services, credentials or components. These joins *allow* whole problems to be solved.

The diagram shows a composite service for a young person moving out of home for the first time. When they register with a doctor there is a handoff to a service that will transfer their mental health provision. When they register for local taxation they are offered the opportunity to apply for secure on-street cycle parking and a new proof of address is created. The new address triggers a notification to confirm the changes to their disability benefit and voter registration. It also identifies potential adult education courses run by the local city government that they may be eligible for.

Some composite services may be very complex indeed, and a user's interactions with them might extend over a long period of time. For example, imagine a family where the parent has become ill and can no longer look after their child, and a relative has applied for guardianship over the child. This is a complex journey that might involve interactions with courts, safeguarding checks, booking meetings with social workers, applying to courts, moving schools, updating benefit claims and applying for social housing. It is unlikely that two families will have the same experience, and the process might take years.[25]

The most important thing in considering the design of a composite service is verifying that each step in the process is 'porous'

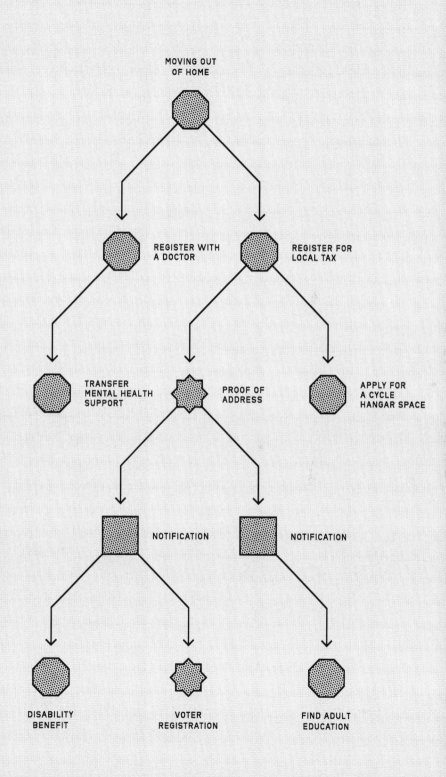

enough to accept handoffs from many other steps. Services that are porous and can be easily joined together create robustness for government because it is simpler to respond to the changing needs of the public when services are not designed to assume a fixed model of how society works.

Composite services differ from abstraction services because they avoid trying to model the whole world in a single user experience. It requires a certain type of humility from policy-makers and designers to design something that is not 'whole', and instead to design joins between services that are open and forgiving.

STRATEGY

Design services that abstract, overlap and join up

Services should be designed around users, not an org chart. But designing public services that collapse institutional boundaries, and enabling the private sector and charities to create complementary services, is a question of incentives. All the design patterns in this chapter rely to some extent on rules and data being exposed as APIs. But because the incentives are typically not evenly distributed, the availability of APIs is not evenly distributed either.

The Making Tax Digital initiative mentioned earlier is an excellent example of public sector rules as APIs. However, those APIs exist primarily because HM Revenue & Customs has a clear incentive to create and maintain them. As a tax department, it has an obvious interest in obtaining better-quality source data for accurate tax calculations and effective enforcement.

New services that cut across organizational boundaries will require access to rules and data. As such, there need to be established processes for the teams creating those new services to request APIs be created. When Argentina was developing its digital driving licence, for example, the National Road Safety Agency was mandated to create an API. The driving licence was then implemented in the miArgentina app, operated by a different part of the government.[26]

Because creating new APIs may not create immediate value for the organization that would need to do the work, there needs to be a presumption that requests will be met. The expectation should be that services will, with very few exceptions, expose their underlying rules as APIs when they receive a request from another service to do so.

Leadership needs to come right from the top. Organizations have to feel they have the permission and support to work in this way. That means leaders have to be comfortable allocating time and money to support work that could have a wider benefit, and they need to make everyone feel it is part of their job to work together. For this reason, one of Amazon's leadership principles, right under 'customer obsession', is an idea of 'ownership' that encourages leaders to 'act on behalf of the entire company, beyond just their own team' and not to 'sacrifice long-term value for short-term results'.[27]

APIs are not enough on their own. Without the right mandate, it's hard to work across the public sector.

Following a 2021 executive order from US president Joe Biden, the executives of nine agencies signed what was termed a 'Life Experience Designation Charter'.[28] This charter established a cross-government priority, along with an interagency team, to improve the 'life experience of recovering from a disaster'. The charter committed the agencies to working together to identify the needs of people recovering from a disaster to inform the service design of a cross-government service: DisasterAssistance. gov. Similarly, Estonia's mission for life-event services came as part of a government-wide strategy.

With the right incentives in place, we might start to see a sort of 'reverse Conway's law' emerging. Rather than services being organized around the shape of the public sector, the work of government needs to organize around services and infrastructure. However, teams tasked with solving complex societal problems must avoid falling into the trap of assuming that every 'life event' can be addressed by a single, new abstraction service. More often than not the answer will lie in better joins.

Main points

1. Conway's law says that organizations design systems that mirror their own internal structures.

2. The public sector has high levels of structural and functional complexity.

3. Complexity creates gaps between services.

4. Services should be designed with the grain of people's lives, rather than against it.

5. Digital can shrink time and space, and networks.

6. Conway's law is felt acutely in the midst of a life event such as the birth of a child.

7. Abstraction services can help users interact with multiple organizations at once.

8. Services that can be loosely joined together as 'composite services' avoid the impossible task of modelling the whole world in a single user experience.

9. Exposing policy rules as APIs means complementary services can be provided by any layer of government and by commercial or third-sector organizations.

10. Public versus private and central versus local are false choices.

11. There need to be mechanisms for services to request the creation of new rules APIs.

12. Organizations need permission and leadership to design services that abstract, overlap and join up.

Examples: anatomyofpublicservices.com/examples/complexity

CHAPTER 3

Certainty over time

There is a paradox at the heart of the next generation of public services: simplification creates complexity.

Services that handle more of the administrative work for the public could radically simplify people's interactions with the state. But more automation will mean more things happening at speed and more things happening at once. This will create new uncertainties for users, and it might mean that important existing safeguards become ineffective.

Humans are not good at multitasking. And it turns out that people who think they are, are not; they just have more confidence.[1] We have limited working memory, and our performance in basic memory tests decreases with load. (For people with dyslexia, ADHD and other neurodevelopmental differences, multitasking can be even harder.)

As the neuroscientist Earl Miller puts it, we live life 'sipping at the outside world through a straw'. Our brains evolved, Miller notes, 'in an environment where new information was usually important'.[2] Computers have increased the production of new information and, once created, it flows faster. Now we live in a world that is filled with red notification dots, push notifications and error messages, lots of it not remotely important, and each a potential attention thief. If they are poorly designed, digital services can end up resembling a 'denial of service attack' on our brains.

Good digital-age design, at least at the level of a single form, web page or notification, is about designing around the limitations of our working memory and other cognitive constraints. This approach is best summarized by the title of Steve Krug's 2000 book: *Don't Make Me Think!*[3]

It's an approach we borrowed when designing the first versions of GOV.UK. Pages on GOV.UK were designed to be functionally readable (easy to scan, and with the available actions easily identified), with the most significant information being most prominent. If necessary we made the font 300% bigger for critical information, such as a tax rate or the date of the next public holiday. We also limited the design of forms to 'one thing per page'.[4] The approach was simple and utilitarian.

Compared with humans, computers are superb at multitasking and at ease with working asynchronously. Complex operations can happen independently of each other, and, for everyday purposes, the availability of working memory is limited only by the budget available to buy more of it.

The designers of public services will more often find themselves having to bridge this mismatch between the varying ability of humans and machines to work asynchronously. It will represent a challenge to the prevailing utilitarian design approaches in the public sector, where the default is to remove and simplify.

Designing for asynchronicity has, in fact, long been a feature of human–computer interactions. The onboard flight software for the Apollo missions was an early example of an asynchronous system, designed so that multiple tasks could happen at once and higher-priority tasks could overtake lower-priority ones as events unfolded.

The user interface for the Apollo Guidance Computer was called the DSKY (short for 'display and keyboard', and pronounced 'DIS-kee'). DSKY used a novel but simple verb–noun interface – 'Fire Rocket', 'Display Time', and so on – to enable a dialogue between the astronauts and the computer for initiating and controlling tasks.[5]

Margaret Hamilton was director of the software engineering team for the Apollo Project at MIT's Instrumentation Laboratory, which was responsible for the development of the onboard flight software used to guide and control the Apollo spacecraft.* Her work at MIT was critical for the success of the moon landings and had a particular focus on ensuring the reliability and error-handling capabilities of the software.[6]

One of the challenges Hamilton and her team had to overcome was representing the asynchronous operation of the software system to the astronauts, especially in the situations when something went wrong. She described the problem in an interview with the Computer History Museum in 2017:

> I started worrying about the astronauts and what ifs, you know. And somehow it worried me, what if there's an emergency and they didn't know it? Because they're just merrily going away, reading the data and putting it in, but what if there's something really major going on and that's it? So I had a meeting with software and hardware people. By software people at the time probably I'm meaning systems people, system designers and everything, and the hardware people. And I wanted to put something in – now, remember, we have an asynchronous environment, right, with all the software. However, we were not asynchronously communicating with the astronauts, okay? We could send something. They'd see the displays, they'd put something in, but we couldn't interrupt their displays.[7]

A new form of user interface was needed to make sense of the asynchronous way the computer was operating and to ensure that the astronauts and the computer could better work together. The solution was to create what were termed 'priority displays' that, in an emergency, would warn astronauts by

* She was also the originator of the term 'software engineer'.

interrupting their displays and giving them emergency options to choose from.[8]

Squeezing an asynchronous design into the Apollo Guidance Computer was an immense achievement given the technical constraints of the 1960s, the need to invent wholly new software engineering approaches, and the context of being blasted into space. Today, software engineers can call on many established patterns for managing asynchronous processes.

'Message queues' maintain lists of current processes and allow the different parts of a system to communicate asynchronously. 'Logs' record the outcomes of different processes running in parallel, while 'promises' and 'futures' represent outputs that may be available at some point in the future. Software implementing these patterns is widely available as open-source products and cloud platforms such as Amazon Web Services and Microsoft Azure. However, as with the Apollo Guidance Computer, services that use these tools to work asynchronously still have users who experience events synchronously, interacting with only one thing at a time. This means people need user interfaces designed to help them understand what is happening and when.

Think again about the complexity of a ride-sharing service. Consider what the user interface has to try and represent as it finds a taxi for you, and then agrees the price of the ride. It needs to communicate the number and location of potential rides and the current level of certainty of the booking, and it has to indicate the likely price. All of these will be the product of a series of real-time negotiations happening in parallel with each other. We could also think of the editing of a shared document, such as a Google Doc, where there is the need to communicate how potentially dozens of users are changing the text in real time and when conflicts arise. Implementing a user interface on top of an asynchronous system is rarely a non-trivial design problem.

The wave of public service digitization that started in the 2000s was mostly about putting paper forms online. A digital form for registering the sale of a piece of land, say, or for applying for or submitting a tax return, just replaced a paper form. Such services do not tend to have much going on behind the scenes, and there is little need to operate asynchronously because they are still, essentially, paper processes. There is still a lengthy back-office process to complete before the user gets an outcome. The data collected by the form progresses step by step by step, like a train moving between stations on a fixed track.

Simple, utilitarian design was a good bet for these sorts of process, but the next generation of public services – more fully designed around the needs of their users and seeking to eliminate as much administrative burden as possible – will have *much* more going on in the background and much more to represent. Users will be able to interact with more than one organization at once, get results in real time, and have automation remove the drudge work of filling in forms (see chapters 1 and 2). These services will no longer operate like that train that moves linearly from station to station or like an office filing system, where a piece of paper can only be in one place at a given time. They will be more like a spacecraft that has new sensor inputs coming in all the time, with events happening in parallel, and undertaking continuous course corrections. Here are a few of the reasons why.

1. An abstraction service designed around a life event such as reporting a death will be initializing multiple tasks in parallel: contacting different agencies such as those for tax and benefits; cancelling credentials such as driving licences; managing a probate application. The exact number of tasks and how long they will take is difficult to know up front.

2. Some tasks will become stuck and need additional information from a user.

3. The once-only principle means a change reported in one service might have significant implications in another service

that were not foreseen by users. For instance, when some-one updates the address on their driving licence, they should understand how this change could affect their benefit pay-ments. Additionally, the process must maintain user agency and accommodate complex situations.

4. The cost of data being wrong will be much higher, because the once-only principle means that if a data record is wrong somewhere, it is wrong everywhere.

5. There is the potential for 'race conditions' (a race condition arises in software when it is trying to do two things at the same time and one task finishes earlier than anticipated).

6. If a digital credential, such as a proof of disability, is revoked or expires, and if multiple services are relying on that creden-tial, how should a user understand the implications?

7. Automating eligibility checks for benefits and grants will remove burdens from users, but they are inherently chal-lenging to explain to users.

The next generation of public services represents a break from the previous paradigm that GOV.UK was designed for, where search was the starting point and written content was the way into services:

search engines → static content → services

The paradigm for the next generation of services is more like a loop. Automation creates tasks to be completed by a user, which in turn feed more automation:

automation → tasks → more automation

Designers will need to consider how and when a service should interrupt a user, when to ask them for additional input and when to seek confirmation. They will also need to decide on a threshold for interrupting a user when the service rules predict there is an additional entitlement the user *may* be eligible for.

Just like the DSKY, tools for navigating the chaos of activity that asynchronous systems create will be needed by the users of public services – tools that help them understand the tasks and events that need their attention. It is helpful to think about these 'tools for understanding' as existing along a timeline:

1. understanding the here and now;

2. understanding what has already happened;

3. understanding what's coming next.

Rather than being stages that a service passes through, they are more like permanent windows onto the past, present and future interactions with a service. Each one provides a clear opportunity to spot and resolve potential issues.

This chapter describes three design patterns: 'journals' for describing the past, 'accounts' for the here and now, and 'task lists' for understanding the future. Each builds on the experience of the design of the Universal Credit service. Journals provide a record of past decisions, events and changes as well as an opportunity to ask for help. Accounts make clear the most important things happening right now. Task lists ensure that it is obvious what needs to happen next, and they also provide a space to show things that may happen in the future. You might also have noticed that these components map nicely onto the asynchronous technologies mentioned above. That's partly because they are the counterparts to logs, message queues and promises.

With tools like these, users can get some of the benefits of automation while also navigating the new complexities they introduce. They are envisaged not as a feature of an individual service but as components that multiple services can make use of. Just as multiple services might use a cross-government content management platform such as GOV.UK to publish information, they would use a cross-government task list or journal.

The design approach towards digital public services since the 2010s has been one of minimalism and simplicity. But the tools described here need interfaces that can flex for radically different contexts, many of them unknowable at the point of design. It's a big change.

In architecture, the utopian minimalism of modernism was replaced with the complexity and contradiction of post-modernism. Post-modernism split buildings into fragments and put unexpected things next to each other. It used ornamentation and symbolism to tell stories, collapse time and set context. The inside became the outside, and angles and colour disrupted the linear.[9]

If it's not too much of a stretch, I'd like to propose that public service design needs a bit less minimalism and a bit more post-modernism. Simplicity will need to mix with transparency and storytelling to help people understand and engage. Interfaces will need to tug at a user's attention, set context and show the workings. They will have to reach into the uncertainty of the future while also providing some certainty about the past.

PATTERNS

1. Accounts describe the now
2. Journals explain the past
3. Task lists reach into the future

Accounts describe the now

If you open a social media app, it shows the latest feed. If you open a banking app, it shows account balances and recent transactions. Music apps show played tunes; exercise apps display progress against goals; mapping apps show your current location.

It has become a common design pattern that the first screen of an app describes the 'state of now'. It's so ubiquitous an expectation that it goes unnoticed (in much the same way that 'save' buttons disappeared when it became an expectation that everything would autosave).

Similarly, a public sector account shows the 'state of now' for a user's interactions with public services. They are a place to surface recommendations and a starting point for important tasks.

The diagram shows an account with 'Claim welfare', 'Driving licence' and 'Local planning'. Each includes glanceable, actionable information to help the user understand what is happening and what they can do next. The welfare claim includes the estimated amount for the next payment. The user's driving licence includes the date it is due to be renewed. The city government's local planning service shows there are two new applications close to the user's home.

Because the user has a child approaching school age, the 'Starting school' service has appeared as a recommendation to add to their account. Over time, we can imagine 'add to account' replacing 'apply for' in how people think about initiating a relationship with a public service. Digital representations of tangible things such as a user's car, a house, a child's school or a company could be added too, grounding people's interactions in the real world (see chapter 7).

Accounts are personalized and contextual spaces. They may look very different the week before a tax return is due or when a house move is underway than they do the rest of the time. A user who has few, sporadic interactions with public

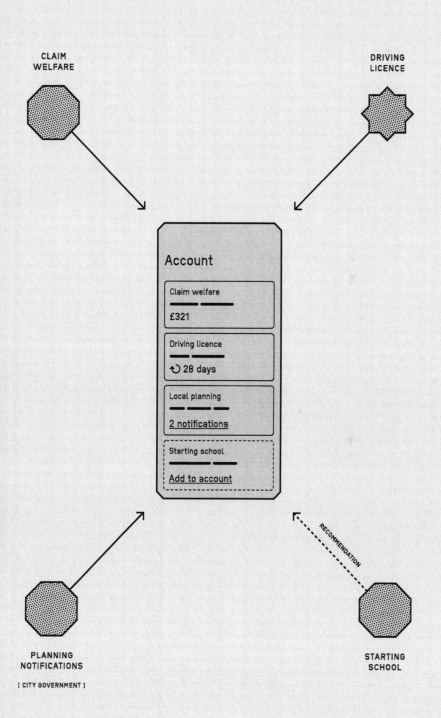

CLAIM
WELFARE

DRIVING
LICENCE

Account

Claim welfare
━━━ ━━━
£321

Driving licence
━━━ ━━━
↩ 28 days

Local planning
━━ ━━ ━━
2 notifications

Starting school
━━━ ━━━
Add to account

RECOMMENDATION

PLANNING
NOTIFICATIONS

STARTING
SCHOOL

[CITY GOVERNMENT]

services will see something very different from someone who is interacting with many services at once, over a prolonged period of time.

Accounts are also a surface on which to engage users in new ways. As well as recommending services that users didn't know they needed, they could provide information that users would not readily seek. As the local planning example shows, that might include opportunities for civic engagement.

Journals explain the past

The next generation of public services will work harder for the public. They will do more, and more will happen at once. Users who rely on those services will need ways to understand what has happened, opportunities to pinpoint things that the service or underlying rules may have got wrong, and routes to resolve any issues.

A journal provides a history of a user's interactions with public services. It is a single space for understanding what happened, why and in what order. Users can view decisions, find important documents and see how data has been used. Journals also provide a clear and consistent route to escalate any concerns.

The diagram shows a journal for a user who has moved house. Their journal is organized around a chronological list of events. In reverse order, they can see that their driving licence has been updated to reflect their new address; the school application process they started is listed; as is the letter confirming that a child's educational needs support will be transferred to the correct new education authority area.

Next, the monthly payment and calculation for their welfare claim are shown, as are the different services that have checked their new address, along with the purposes of each check.

After moving house, they find they have a problem with their benefit payments. Using their journal they were able to identify that the monthly calculation had not taken account of their new

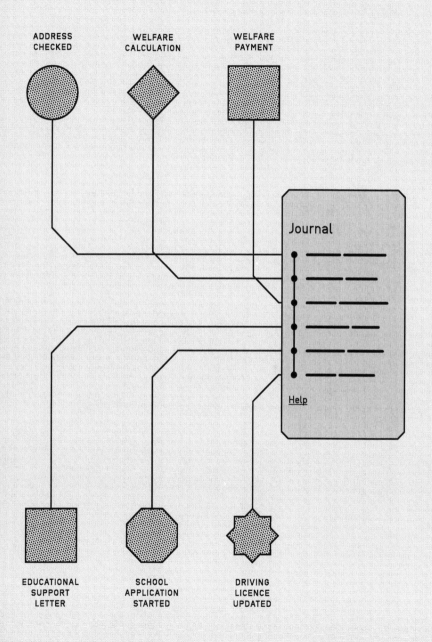

address, meaning that they were underpaid. They submitted a request for help to resolve the issue.

A journal is a shared surface for users, advisors and government officials to understand and resolve issues. In situations where government officials are unable or unwilling to resolve an issue, a journal provides proof that a user can use to explain their situation to others who can and are willing to help.*

Journals also provide part of the answer for the automation of decision making in public services. They provide both a unified design approach for surfacing automated and manual processes and a place to link to appeals and support.

This type of audit log has been impossible to date. Records are incomplete and split across emails, text messages and letters. Digital services have often entrenched this problem, replacing letters one for one with emails and notifications without thinking about how everything joins up.

Given the richness of information included in a journal, and the detailed view it could give of aspects of someone's life, they represent an information security, privacy and design challenge. These challenges will need to be front of mind for anyone designing and implementing a journal.

In time journal-like components will become a precondition for public automation. Not only do they provide a space for explaining and understanding, they place an obligation on public services to put on record what happened and why.

Task lists reach into the future

Task lists make it clear what a user has to do next. They give a single view across multiple services, and they provide a unified design approach for surfacing automated and manual processes.

---

* Stephen Timms, a member of the UK parliament, has said that constituents sometimes come into surgeries with their Universal Credit journal loaded ready on their phone so they can point out mistakes.

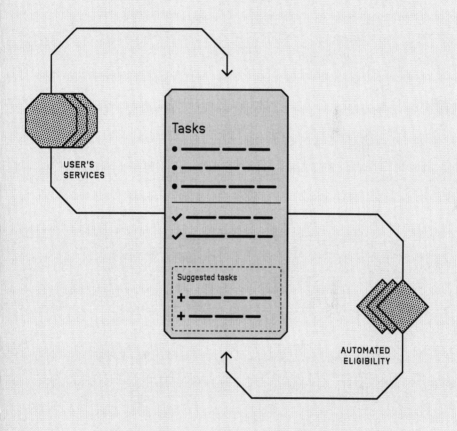

The diagram shows a cross-government task list and illustrates the two ways tasks can be added.

First, services can add tasks. For example, if a parent is applying for an 'educational needs assessment' for their child, that service might add tasks to upload a diagnosis letter, report the child's details and share details of their school.

Second, automated eligibility rules can suggest tasks. So, if the application for the educational needs assessment was successful, automated eligibility checks might recommend a task to apply for disability payments or to review local support for children with special educational needs.

Task lists can also surface work that is being done by services on behalf of a user (see chapter 7). For example, a user reporting a health condition might see a list of tasks being completed by government agencies on their behalf. Similarly, a user applying for a visa might see that their paper documents are pending a check by an official.

Task lists place a responsibility on services to tell a user everything they need to do. This is very different from the model of legacy services that place an unreasonable responsibility on users to report any 'change of circumstances'.

STRATEGY

Unbundle services

It's surprising how few public services tell people what is happening now and what will happen next, or provide a record of past interactions. New 'tools for understanding' could change that. It might simply be that they don't have a place to surface this information effectively. It might also be a product of how digital services are designed and built as fixed, tightly integrated, end-to-end user experiences.

Accounts, journals and task lists should be operated as common components that every public service can use. As with single government websites such as GOV.UK, they need to be funded and maintained for the whole public sector. Additionally, they must be designed to make it as simple as possible for services to publish information to them.

But once they exist, how do you fill them with enough useful stuff? The answer lies in 'unbundling'. Unbundling is the breaking down of a user experience into its component parts so that those parts can be remixed, reused and reassembled.

Today, tasks are locked away in long digital forms. The raw materials for journals and accounts are scattered across notifications, emails and letters. Just as Apple has done with its Shortcuts app (see chapter 2), the public sector needs a strategy to 'unbundle' the common aspects of government services into their component parts so that they can be reassembled as features of task lists, journals and accounts.

Common components such as notification systems and form builders have a role to play here too. As well as sending a notification, a notification component could also update a journal. A forms component could support tasks being added to a task list.

Part of the shift described in this chapter is the move from 'the service' as the organizing unit for a user's interactions to 'events' and 'tasks' being the organizing units. Without an unbundling strategy, governments who are redesigning services one by one risk baking-in an interaction pattern that is already out of date and may be costly to unpick.

Main points

1. Paradoxically, simplifying services for users though automation can introduce new complexities for them.

2. Computers are good at multitasking, humans are not.

3. The new complexities are the result of real-time interactions, more things happening in parallel, more data being reused between services, and the abstraction of organizations.

4. Users need new tools to navigate these complexities.

5. Journals can help users to understand what has happened in the past and to verify outcomes.

6. Journals give users something they can use to explain their situation to people who are helping them.

7. Accounts give users a clear understanding of the current state of things: which credentials they can use and which services they are interacting with.

8. Task lists help people understand what they need to do next and to surface recommendations.

9. Services need to be 'unbundled' to create the raw materials for journals, accounts and task lists.

10. Accounts, journals and task lists represent surfaces for automation and prediction to scribble upon.

Examples: anatomyofpublicservices.com/examples/certainty

CHAPTER 4

Shards of identity

In his book *A Mad World, My Masters*, BBC journalist John Simpson tells how, in 1976, he travelled from Africa to Europe for a month on a 'passport' that was little more than a scrap of paper. Having had his British passport tossed in a river by a soldier, the part-time honorary consul of the Netherlands in Angola (apparently the sole representative of a European Economic Community country) had hastily typed, signed and stamped a letter for him. The only personal details it included were his name and his employer. It became so creased and dogeared that it was almost unreadable, but people trusted it enough to let him cross borders.[1] Just as with money and debt, a credential is, by definition, something people trust. Otherwise, it's just a bit of paper.

If you want to prove that your child has a disability in the United Kingdom in 2024, you carry a similarly dogeared bit of paper around. Parents have to use a printed 'award letter' for a disability benefit to gain access to disability facilities. If, for example, an autistic child needs to jump the queue at an art gallery, then the letter is unfolded at the ticket desk and an explanation is given to anyone within earshot. There are whole Facebook groups dedicated to discussing which amusement parks and visitor attractions will accept the letter and which will not.

Where the award letter differs from Simpson's 'passport' is that it includes a lot more information. The letter's intended purpose is to tell a family what financial support it is getting and why, not as a proof of disability. As such, it includes a break-down of the child's needs (such as 'help with personal care'), the money awarded for each of those needs, and the child's address – more information than is strictly necessary to access the Tower of London. It has become a de facto credential in the absence of anything better.

These sorts of de facto credentials are common in the United Kingdom. Proving your address requires a bank statement, a phone bill (landline only) or a local government tax statement. That means you have a choice between sharing your bank balance, divulging who you phone, or revealing the value of your home. Since most banks have now digitized their statements, and with fewer people even owning the things we used to refer to as phones, a paper tax letter is the only option for many people.

Even Universal Credit, a modern digital public service, instructs its users to download a screenshot of their digital statement (which includes a breakdown of how much money they receive and why) as proof they are getting the benefit.[2]

The United Kingdom might be an outlier in quite how reliant it is on random bits of paper due to its lack of a singular popula-tion register or identity document. But wherever you are in the world, physical identity documents tend to reveal more about their holders than the task at hand generally requires.

Digital identity is different. In a 1998 paper with the very 1990s title 'Digital identity in cyberspace', Hal Abelson and Law-rence Lessig set out why.

First, digital identities can be treated as a set of individual attributes rather than a single homogeneous lump. So, some-one can prove that they have a disability without revealing the details of it, or they can prove their address without reveal-ing their bank balance. Second, digital identities can also be 'detached' from someone's real-world identity. Physical identity documents are presented in person. In John Simpson's case, it no doubt helped that his 'passport' was presented along with his physical self – a white British man with a private school accent who had been on TV a little. (Simpson also noted that British imperialism meant that the defining characteristic of the British abroad was that they carried no identity documents until the early twentieth century.) So, the digital identity someone uses to submit a tax return may be strongly bound to their real-world identity, but the one they use to do crosswords online may have little or no link at all. By implication, this also means someone can have more than one digital identity.

Abelson and Lessig called these two features of digital identity – the fracturing into individual attributes, and the degree of separation from a real-world identity – 'unbundling'. The degree to which a digital identity is unbundled is, they said, a political and an architectural choice.

The ability to unbundle digital identities is useful in the context of public services. Varying the degree to which someone's identity can be traced back to them means that someone can create a short-lived digital identity for a specific purpose. For example, they could prove they had a valid driving licence at the point when they hired a car without including a link to their full personal details, essentially eliminating the risk of that digital identity being used by someone else at a later date.

The Indian digital identity system Aadhaar is linked to biometrics, so it has a very strong link back to a person's real-world identity. But users can also generate a 'virtual ID' number that they can share more widely and that can be regenerated at any time. They can also generate share codes that give one-time access to information. Similarly, Ukraine's digital passport, which is used extensively in everyday life, means that people no longer have to present their physical passport. The government's Diia app generates a temporary QR code every three minutes, so the digital identity that someone uses to buy a SIM card is different to the one used to buy a train ticket.[3]

Digital identities also enable people to compartmentalize their lives, with a separate digital identity for health matters and their personal interactions with government, as is the case with the United Kingdom's NHS login (health) and GOV.UK One Login (government).

Finally, the unbundling of attributes means that only the minimum amount of information that is needed for any given transaction is shared. If all a service needs to know is that someone is a resident of Massachusetts, or claims food stamps, or has children, then that's all that needs to be revealed. Rather than sharing everything about someone, it is possible to share facts (e.g. 'over 18', 'registered disabled', 'earns less than median wage'). Digital identity can become less about 'I am' and more about 'I have done'. These facts are the basis of digital credentials. They are shards of identity, or 'tear-off strips' if you prefer, containing just enough information.

Unlike in the 1990s, when 'Digital identity in cyberspace' was written, digital credentials are now a reality at scale. As of the beginning of 2023, 80% of drivers in New South Wales had a digital version of their driving licence. France, Malaysia, Argentina, Norway and Brazil have all launched one too.[4] While in British Columbia lawyers can access court services and documents using their 'Law Society of BC' credentials, which are held in a government app.[5]

The largest-scale implementation of credentials is based around DigiLocker, the Indian government's credentials app. It has 146 million users and has issued 5.6 billion credentials for everything from pensions to birth certificates. The ecosystem continues to grow, and India's 2023 budget contained an announcement for a version of DigiLocker for companies and charities to manage their credentials too.[6]

While India's approach to digitizing credentials relies on a publicly owned app for storing and sharing them, many US states now issue digital driving licences via the Apple Wallet and Google Wallet apps.[7] In addition to digitizing credentials, Apple is also unbundling them. Third-party iPhone apps can request a specific bit of information, derived from the credentials: that someone is over 21, for example.[8]

Where credentials will 'live' is both a technical question and a political question. Apple's and Google's digital wallets, and those of Samsung and others, are turning the storage of credentials into a zone of contest between the public and private sectors. With DigiLocker, India has claimed credential management as part of the public sector, and the EU and its member states have done the same through their EU Digital Identity Wallet.[9]

There are different models for credentials storage emerging from a technical perspective too. India's DigiLocker takes a more centralized approach, providing a 'shareable private space on a public cloud' with 'authentication, consent, audits, and other security best practices'.[10] South Korea also follows the centralized wallet model to store things such as digital versions of students' degree certificates and proof of residence via the National Digital Document Wallet.[11]

Canadian provinces and the EU are, by contrast, exploring a model that uses a system of decentralized identifiers, or DIDs, stored on a blockchain or other distributed ledger to record which credentials have been issued. The credentials are stored in an app on a user's device and can be used without accessing a central database. Decentralized credentials do not replace governments as issuing authorities (a common misconception – a digital driving licence not vouched for by a government agency is not very useful!), and governments still need to maintain their records of what credentials have been issued and why. Decentralized credentials are better thought of as widely trusted vouchers that represent a record held somewhere in a government database.

There are tradeoffs to be weighed between centralized and decentralized models. Decentralization promises greater privacy for users because credentials can be used without government knowledge. So if someone uses their government-issued address credential to get a home insurance quote or order a pizza there is (theoretically at least) no central record of those interactions. On the other hand, centralization may create privacy concerns but also make it easier to spot fraudulent use of credentials and inform users when data about them has been accessed.

The hard thing for the public sector to navigate in making some of these decisions is not so much what credentials are today, but what they might *become*. Today, digital credentials are presented as being relatively dumb, living in inert spaces (wallets). But they have the potential to be more than trusted facts, and if the number of credentials issued by the public sector soars, they could take on some very different qualities.

In addition to digitizing state identity cards and driving licences, Apple and Google are also working with the US Transportation Security Administration (TSA) to design an end-to-end process for people to use those identities to board internal flights. When using the service, a user presents their device at a custom-built government kiosk that contains a screen and a bluetooth reader. The user's phone then displays their credential and explains what information will be accessed by the TSA. The user then confirms they understand and can proceed to board the flight.[12] It is the credential that starts the process and mediates the interaction.

The TSA example shows how credentials are the raw materials for automating eligibility checks within other services – eligibility to board an aircraft in this case. It also starts to show how credentials can be more than trusted facts. They can be a way into services and an object to organize other information and actions around. They can be what are termed 'social objects'. '

The idea of social objects in the design of digital services was proposed in 2005, in the heyday of 'Web 2.0', by Jyri Engeström, a Finnish entrepreneur who would go on to co-found Jaiku, a microblogging service that predated Twitter by a month.[13] In a blog post titled 'Why some social network services work and others don't', Engeström rejected the idea that successful digital social networks are formed between people, instead arguing that they are based on 'object-centred sociality'.[14] Rather than thinking of social networks as being made up of people, he said we should consider them to be networks of people connected by shared objects. Digital social objects could be anything digital that held meaning or interest for people: an event, a URL, an image. He saw such objects as a placeholder for *activity*.

The social object concept was based on the work of the sociologist Karin Knorr Cetina, who studied how knowledge-based communities created knowledge. She found that, for scientists, instruments such as microscopes and particle accelerators are not just used for observations, they are actively involved in defining what questions are asked and what experiments are conducted. Studying global financial traders, she found that financial instruments such as derivatives and trading algorithms don't just facilitate transactions, they shape market behaviours and decisions and are continuously modified in response to market conditions. These objects, she said, take on a social form within those communities, 'like open drawers filled with folders extending indefinitely into the depth of a dark closet'. They continually acquire 'new properties and change the ones they have', and they are 'as much defined by what they are not (but will, at some point have become) than by what they are'.[15]

At the time of Engeström's blog post, the go-to reference for social objects was the photo-sharing website Flickr.* On Flickr, social objects were the photographs that were shared on the platform. It was around those photos that communities formed, discussions happened and relationships were created. The photos were not just images, they were catalysts for activities such as commenting, tagging and sharing across other networks.[16]

The category error that Engeström identified in 2005 – that social networks are not merely agglomerations of people – has broadly proved itself true. Even if many of the Web 2.0 wave of social media services have been eclipsed by the likes of Facebook, social networks do form around social objects, be they videos on TikTok or running routes on Strava.

---

* Facebook would not be available to the general public until the end of 2006.

Today, a similar category error to the one Engeström identified in social networks is at play in digital public services. It is common to think of a relationship existing between a user and a service. But this is to put the relationship on the government's terms, and, in turn, it frames digital identity solely as a means to access public services. Instead, we should see the relationship as existing between a service, a user and a *credential*. It is the credential that is the starting point: the object that the relationship forms around.

If a driver receives 'points' for unsafe driving, those points are 'added to the driver's licence'. The number printed on a driving licence might be required to pay a fine. The licence is the object that mediates the relationship between the driver and the police, and the driver and the courts. Credentials provide the context for an interaction: when someone uses a company registration credential, they are acting in their capacity as a company director not their capacity as a school governor or the driver of a vehicle. Finally, credentials describe a set of privileges, powers, rights and duties between the state and an individual or organization.[17] A successful claim for a tax credit places a duty on government to honour that claim. A government agency may have the power to revoke a licence to abstract water from a river, while a utility company may have the power to extend a licence to dig up a road. A valid driving licence grants someone the privilege to drive (within the law). If they break the law, they may have to pay a fine or other penalty.

Credentials are social objects: the manifestation of the relationship between a service and its users and of the different rights, duties and privileges that each side enjoys.

Just like the photos on Flickr, digital credentials can surface activity and actions. They can have an 'affordance of interaction'. Credentials can be updated to reflect changes in the relationship between the user and the service that issued the credential, becoming surfaces for proactive services to scribble on. They can be a starting point for tasks such as making payments, completing tasks or checking how data has been accessed.

Credentials as social objects represent a break from the model of government services being organized around concepts such as 'apply' or 'renew'. With social object thinking, those concepts just become features of a credential rather than services in their own right. The credential is something real – something tangible that represents the relationship and what a user can do with it.

What we don't yet know is how credentials might change the nature of the relationships they mediate once they are digital. Does the creation of more verifiable data about users mean that hyper-means-tested services like Universal Credit become more common because they become easier to create? Will more credentials mean people have their credentials checked more often, just because that is possible? Will the 'power to check' go to the heads of policymakers?

Certainly, the widespread use of digital credentials will need some new norms and regulations. The launch of the state of Louisiana's digital driving licence, for example, was accompanied by new laws that prevent the police from taking people's phones off them and looking things up when checking a licence.[18] The right to use digital credentials must be balanced by a duty of the public sector to use them responsibly. How those responsibilities and rights are framed though, matters.

The common framing for such discussions is privacy, data minimization and user control. Privacy is something that is preserved by limiting access to information. Privacy debates tend to attract absolutists on both sides, with sometimes-arbitrary arguments that everything must be put under user control in the name of privacy, or the counterargument: that it doesn't matter what information is reused because people assume the government knows it anyway. Both are unhelpful.

Public services, by their nature, often deal with incredibly sensitive information. The fact that digital identities can be unbundled into multiple attributes and separated from a real-world identity should not be seen, in the first instance, as an opportunity for data minimization. That is often a false choice anyway: if someone requires financial help, asking them to surrender information to get that help is not really a choice. If the law demands it, it is even less of a choice.

Lisa Austin, chair in law and technology at the University of Toronto, points to an alternative way of thinking about identity that is more helpful: shame. Shame is not how we appear to others, or whether something is shameful, or private, in principle. Shame is how *we feel* about how others *see us*. It is self-reflective, constructed through social interactions with other people.[19] Through shame, Austin recasts privacy as something that is less about limiting access to information and more about enabling relationships by allowing people to present the version of themselves they want people to see in a way that minimizes shame.

Digital credentials, conceived of as social objects and taking full advantage of the unique properties of digital identities, should be considered as a way to enable people to construct relationships with the state and other institutions that minimize shame by giving them control over what they present and how.

PATTERNS

1. Credentials are the output of services and inputs into others
2. Credentials are the raw materials for automating eligibility
3. Credentials are social objects

Credentials are the output of services and inputs into others

Digital credentials are alternatives to the letters, certificates and other physical documents that people use to prove their identity, their entitlements and their rights.

Services represent opportunities to create credentials. When a user makes a successful claim, completes a registration or files an application, there is an opportunity to issue a credential. Moving house can create a new proof of address credential. Registering a company can create a company registration credential. Claiming benefits can create a proof of benefits credential or a proof of disability credential.

Credentials are digitally signed by the issuer, so they can be trusted by other services that need to use them. This can be a starting point for a service: where one entitlement entitles someone to claim something else. A disability parking permit may entitle someone to a disabled parking bay outside their house, for example. Alternately, credentials can be used in combination to prove eligibility for a service. A visa renewal service may need proof of citizenship, proof of address and proof of income, for example.

The diagram shows a credential that is created when a child leaves the care system through adoption. This *previously looked after child* status unlocks access to further support both as a child and as a young adult. In the short term it is used by the adoptive parents to access priority education places and funding, and to claim social security means-testing exceptions. In the longer term the credential passes to the young person, who can use it to access additional support when applying for a place in further education, such as support with accommodation and additional funding. Because the services consuming the credential just need to know that it is valid, the young person does not have to explain their background to strangers if they don't want to.

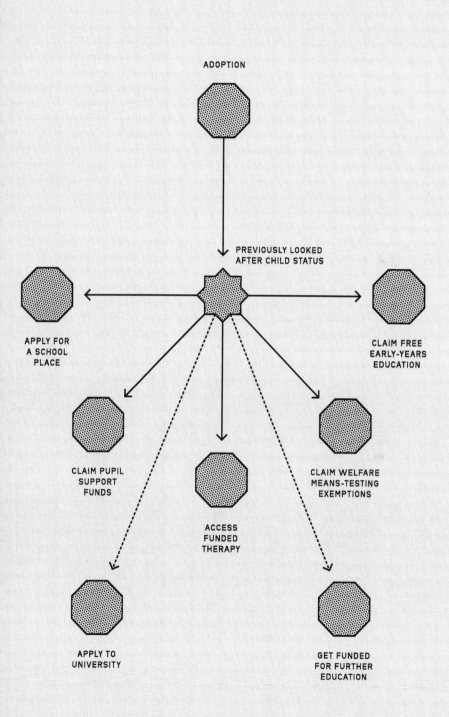

ADOPTION

PREVIOUSLY LOOKED
AFTER CHILD STATUS

APPLY FOR
A SCHOOL
PLACE

CLAIM FREE
EARLY-YEARS
EDUCATION

CLAIM PUPIL
SUPPORT
FUNDS

ACCESS
FUNDED
THERAPY

CLAIM WELFARE
MEANS-TESTING
EXEMPTIONS

APPLY TO
UNIVERSITY

GET FUNDED
FOR FURTHER
EDUCATION

It's not possible to know all the ways in which a credential might be used. Once they are in the hands of users, credentials create new service interactions and policy options. To continue with the example above, a future policy aimed at previously looked after children becomes easier to implement if a trusted, secure credential exists.

Credentials are the raw materials for automating eligibility

It is impractical to create real-time, proactive services based on data that users have entered themselves. Automating eligibility for things such as benefits and visas is dependent on the availability of trusted data.

Credentials are, in their simplest form, bits of trusted data that represent things such as an entitlement, a characteristic or a right. Because they are digitally signed by the issuer, they can be verified before use.

The diagram shows someone using a conversational AI. The user is in the process of renovating their house and wants to know what support they might be eligible for to reduce their bills. With permission, the system has accessed their credentials and uses their address, their income and the energy performance certificate of their house to check for grants. By comparing the credentials against the eligibility rules for different programmes, the AI not only identifies potential grants, but also uses those credentials to start applying for them.

Digital credentials allow for some radically different interaction models for public services. As in this example, users may not interact with the service directly at all. Digital credentials can mediate interactions with services while simultaneously helping a user understand what information is being used. They change the balance of ownership between the user and the service: credentials are the outputs of services, but they are owned by users.

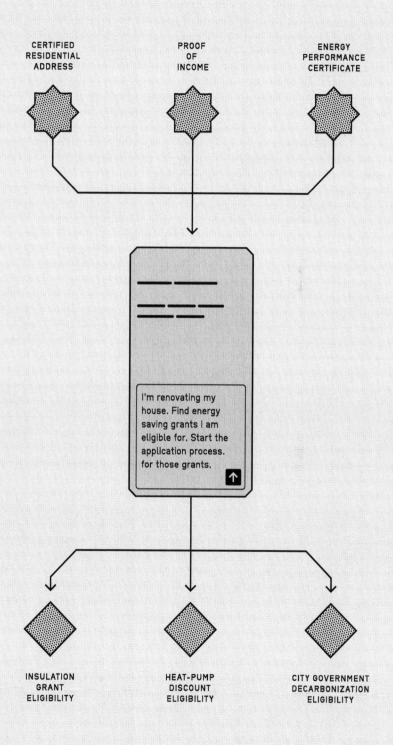

CERTIFIED RESIDENTIAL ADDRESS

PROOF OF INCOME

ENERGY PERFORMANCE CERTIFICATE

I'm renovating my house. Find energy saving grants I am eligible for. Start the application process. for those grants.

INSULATION GRANT ELIGIBILITY

HEAT-PUMP DISCOUNT ELIGIBILITY

CITY GOVERNMENT DECARBONIZATION ELIGIBILITY

Credentials are social objects

Credentials have *salience* and *potential*. They can surface activity and actions, and they mediate interactions.

The diagram shows a driving licence credential. Services can request access to it as part of a digital interaction, the QR code can be scanned in person, and a user can generate a code for sharing information over the phone.

The driving licence also serves as a space to surface related actions and information. The current status of the licence is shown – in this case it will auto-renew in one month – which allows the user to manage the renewal process. (For example, if they will soon be driving abroad, they may wish to renew it sooner.)

Three services have recently checked this licence so these are listed next to it. If the user is curious about who has been checking it, they can click through to see the access history in their journal (see chapter 3).

Details of the user's car are also shown, as are tasks related to it. The user can verify that their car is taxed, pay a city congestion charge or manage a transfer of ownership if they are in the process of selling it.

If a user updates their residential address in another service, such as when completing a tax return, showing the update on the credential is a way of explaining that the process has completed.

In these ways, credentials have the potential to shift the starting point for public services interactions from abstract processes to something more tangible (see the concept of psychological ownership in chapter 7).

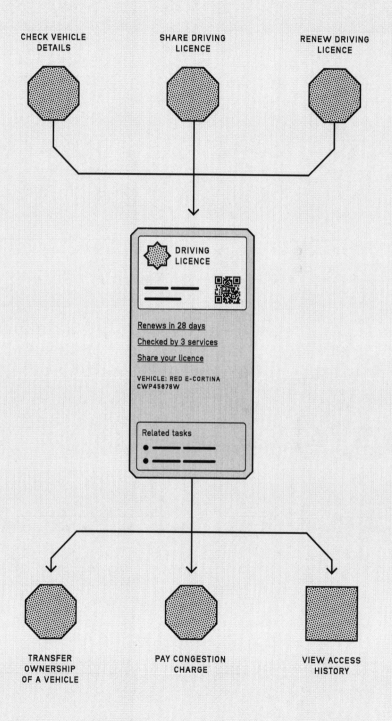

Digitize all credentials

In many countries, digital driving licences have proved the case for digital credentials. But there's a risk that, as with physical credential documents, only a small number of high-value credentials are created. A future in which digital driving licences exist but where people still have to show printouts of emails and screenshots to prove they get disability benefit, are eligible for free school meals or were looked after in the care system is a missed opportunity. People should have a right to digital versions of all credentials. That will mean getting the incentives right.

Creating 'proper' paper certificates or entitlement cards rarely created value for the organizations best placed to create them. Things such as paper benefit letters became de facto credentials because there was no value for the organization that printed them to do anything more. Benefits agencies send a letter saying someone meets the eligibility requirements for a disability benefit because they must tell them the outcome of an application. The letters include a registration number to make it easier for the agency to find the recipient's records when they phone up. But the agency has little incentive to make it easier for someone to claim additional support from a city government or to access a visitor attraction. The value of credentials is created when other services consume them.

Realizing the potential of digital credentials will require something more systematic: political will and digital infrastructure.

First, there needs to be a clear mandate for all services to prioritize the work of creating digital credentials. The French government has begun to employ this exact strategy, which it terms dématérialisation. Ukraine also talks of 'dematerializing documents',[20] while DigiLocker's mission is 'eliminating the need for physical documents'.[21]

There should be an onus on services to understand how their letters and certificates operate as de facto credentials. Paper documents that cannot prove they have no secondary use should be digitized.

Bold changes like these might seem unobtainable, especially for states that are still heavily reliant on paper, but the systematic digitization of paper is a more achievable aim than service-by-service digital transformation.

A 2016 law put documents stored in DigiLocker on a par with original physical documents,[22] while Australia legislated in 2023 to end the need for paper-based signatures when the rules around digital signatures and video witnessing introduced during the Covid-19 pandemic were made permanent.[23] Similar laws could make it a requirement that all services issue digital versions of their credentials.

This does not mean that paper versions should cease to exist – the United Kingdom's digital-only immigration status shows the risks of discrimination from fully eliminating physical documents[24] – but people should be able to access digital versions if they want them.

Second, it is much easier for services to issue and check credentials if there are components they can use to create and check them. This is what the government of British Columbia has done. Their Digital Trust team has created components that services can use to issue and accept credentials. These include an app for storing credentials and a verifiable data registry for revoking them.[25]

Similarly, part of the adoption strategy for India's DigiLocker app is to work with whatever legacy systems already issue. Rather than insisting on the full digitization of existing processes, government agencies can, as a first step, choose to make existing paper documents available in DigiLocker, as what is essentially a PDF. More modern services can decide to issue fully verifiable digital documents.

Creating components that make it easier for services to work with credentials will require a better understanding of what good interactions look like for issuing and using a digital credential. In the early days of the web there were plenty of examples of clunky sign-in forms. But today, if you ask a designer to create an interface for someone to sign in with a username and password, manage a password reset and set up two-factor authentication, they will likely do a pretty good job of it. This is because there are now many good examples out there.

Credentials are currently in their early days. If you look at the process for using credentials from DigiLocker as part of a loan application, for example, it is pretty clunky. Handling the failure cases where credentials are out of date, or where a user elects to use a paper credential instead, is something that will need particular attention.

The idea of credentials living in a 'wallet' feels pretty sub-optimal too. Having a specific app for credentials seems technocratic and may limit the opportunities for credentials to be designed as social objects that mediate wider interactions with services. Credentials must become part of the design language of government services, not be siloed off into a wallet (be that a public sector or a private sector one).

Bringing credentials closer – at least in design terms – to digital accounts feels like it would open up more opportunities. There are some examples of just this. The government of Western Australia has made digital credentials for a seniors card, recreational boat licences and fishing licences available in its 'Service WA' app. And in Argentina, the country's digital driving licence is positioned within the miArgentina app.[26]

Main points

1. Letters and other documents become de facto credentials in the absence of anything better.

2. Physical credentials reveal more information than needed.

3. A digital identity can be strongly or loosely bound to a real-world identity.

4. A digital identity can be split up into attributes.

5. Shame is how we feel about how others see us.

6. Where credentials will 'live' is a contested question.

7. Digital identity can become less about 'I am' and more about 'I have done'.

8. Digital social objects could be anything digital that holds meaning or interest for people.

9. Credentials are social objects.

10. Credentials are the raw materials for automated eligibility checks.

11. Credentials are the outcomes of services and the inputs to others.

12. Every credential should be available as a digital credential.

13. Common components can make it simpler for services to issue credentials.

Examples: anatomyofpublicservices.com/examples/shards

CHAPTER 5

Common components

If you walked into a government office in the twentieth century, you'd find stacks and stacks of paper. On the outside of the building you'd find a sign that said something like 'driving', 'pensions' or 'tax' and the logo of a government agency with top-to-bottom control over that policy area.

Forms would come in via a post room and then be sent on to the correct administrative team. That team would be busy validating forms and managing casework according to policy rules and legislation. Processed cases would end up in the records department as some mix of paper files, index cards, microfiche and – later in the century – files on mainframe computers. More paper would come out of the other end as letters, licences and certificates. In 1999, when I was a scientific officer at the UK Veterinary Medicines Directorate, half of my office was taken up with racks of files and by the giant 'burn bag' in the corner (this was for disposing of confidential documents and was emptied once a day) while the bulk of the ground floor was set aside for the storage of paper records.

For larger government agencies the amount of paper they processed was truly vast. The US Internal Revenue Service (IRS) created dedicated zip codes for different groups of taxpayers. Zip code 73301, for example, is for international taxpayers, and it automatically routes mail to an IRS office in Austin, Texas.

Similarly, the United Kingdom's driving licensing, pensions and tax departments handled so much paper they were given dedicated postcode districts (a district is normally reserved for a whole town). In 1978 the Driving and Vehicle Licence Centre in Swansea (postcode district SA99) alone was handling 12% of all the mail in Wales. It had twelve postcodes, one for each department. The data collected on the paper forms was destined to end up on magnetic tape, processed by three mainframe computers in a giant computing room. The driving licence file took up 106 reels of the stuff, and the vehicle file took up 134.[1]

As governments around the world began to put their services online, they digitized this world of paper. Paper forms were transformed one-for-one into digital forms. Case management systems – which were generally digitized earlier than public-facing services – encoded policy rules in bespoke systems that automated things such as validating addresses, managing payments, verifying someone's identity and printing letters.

Digitizing a world of paper created a level of 'technical debt' that many governments are still trying to unpick to this day. Just as a bank loan means that money is released to an organization so that it can use it to buy something sooner but will have to pay interest over time, technical debt is created when an organization does something for short-term gain that stores up work for the future.[2]

If you were to look at any three government agencies – driver licensing, land registration and tax, for example — you'd find that their systems did some very similar things. They all need to create digital forms, validate addresses, manage payments, verify people's identities and print letters. They all need servers to host their systems on and databases to store their data in. What is more, the costs of operating three payment systems and three letter printing systems rather than a single one are higher. Public money therefore gets wasted on doing the same thing in multiple different ways.

In 2016 the UK Government Digital Service tried to get to grips with the scale of the duplication. It surveyed 150 teams across the United Kingdom's central government and found that 85% needed to send a notification to their users, 30% needed to verify users' identities, 65% had to have some sort of two-way conversation, and 80% had to collect information from their users. Mostly, those teams had to create their own, bespoke ways of doing those things.[3]

Complex systems are also harder to build and to change. Every time a new service is created it takes longer because the teams creating it have to 'reinvent the wheel', hand-crafting systems for authenticating a user, hosting a database or publishing a map. Maintaining a system that issues driving licences that *also* needs to verify someone's identity and accept payments is just harder.

As well as giving rise to technical debt, digitizing a world of paper created new administrative burdens for users. It's more confusing for users if the design of a payment receipt is different, if each form has a different design, and if people have to go to a new place to read their messages.

Concerns about digitization often focus on people without digital skills. The latent cost of services that have never been digitized is under-debated. It has typically been unrealistic to expect smaller agencies to take online payments, provide online support or create a digital form. That means low-volume services remained stuck in digital limbo, with their forms published as PDFs.

The high cost of creating services meant that, even for larger agencies, digitizing their forms was typically a piecemeal process, with only the most-used forms being digitized and the rest turned into PDFs.

As of 2024 the IRS lists more than 800 different forms on its website. The United Kingdom's driving agency, now named the Driver and Vehicle Licensing Agency (DVLA), still receives 60,000 items of mail a day (down from 80,000 at the turn of the century).[4]

During Covid-19, the IRS ended up with millions of items of unopened mail, including tax returns for international students, which could not be submitted digitally. Unopened mail ended up stacked in containers in the car park at the IRS facility in Austin, Texas, and boxes of paper returns filled the staff canteen.[5] The DVLA also got hit by a considerable backlog of paper forms and physical driving licences.[6] The stress for civil servants and the public must have been enormous.

Conducting digital transformation service-by-service, form-by-form, one casework system at a time is not the answer. It might be great for consultants, but it will mean that the heat-death of the universe will arrive sooner than every service being brought up to 2005 standards.

In 2011 Tim O'Reilly published an article called 'Government as a platform' that pointed to a different approach. One of the questions O'Reilly posed was: what if governments were organized more like operating systems, citing the Unix operating system as a model.[7]

Unix, created in the 1970s, paved the way for modern computer operating systems such as Linux, Android and macOS, which follow Unix's design principles of simplicity, modularity and composability. Those design principles are best summed up in the Unix philosophy:

> Write programs that do one thing and do it well.
> Write programs to work together.
> Write programs to handle text streams, because that is a universal interface.[8]

In Unix-like systems, the job of the 'find' command is to search for types of file and that of the 'wc' command is to count stuff. Those commands can be combined to count the number of PDF files in a folder. If the 'id' command is added, it counts the number of PDFs owned by the signed-in user. Each command does one thing well and can be combined to create something that is greater than the sum of its parts.

We see this pattern of componentization in many digital systems. So whether you are watching a video or ordering a taxi, the system you're using is likely to be built from some of the same parts, operated by software-as-a-service and platform-as-a-service providers such as Amazon Web Services (AWS), Microsoft Azure, Stripe and Twillo.

The likes of Amazon and Stripe make money because they make it faster and simpler to create services. Their components do one thing well: send an email, authenticate a user, accept a payment. What they do is clear, as is how to use them (many of AWS's offerings have the word 'simple' in their name). They are designed around the needs of software engineers and can be swapped in and out without affecting other parts of a service.

In the decade and a half since O'Reilly's premonition, what was obvious to the designers of Unix fifty years ago has become a way for nation states to think about their national infrastructure: computer systems are built from small parts, loosely joined.

Around the world, governments are building and operating common components. For some, it has even become an element of their foreign policy.

In the United Kingdom, GOV.UK started providing a common way for publishing written content in 2012. GOV.UK Pay was added in 2015 to handle payments from the public, and the GOV.UK Design System – a set of tools for designing constant user interfaces – was added in 2016.[9] GOV.UK Notify was created to enable services to send emails, physical letters* and text messages.[10]

The GOV.UK components are used by services and organizations across both central and local government, mostly for free. They are paid for and operated for the benefit of the whole public sector. At the end of 2023 there were more than 1,400 organizations using GOV.UK Notify, many of them small government agencies, schools and healthcare providers.

Other governments have similar programmes. In 2016 Italy hired an Amazon executive, Diego Piacentini, to set up a central

---

* The letter printing function used spare capacity at the DVLA and its excellent integration with the postal system.

digital team. It followed an approach very similar to the Amazon Web Services model, creating components for payments, digital identity and a design system for mobile apps.[11]

The government of British Columbia has built components for accepting payments, creating maps and scanning documents. It has also recently created a common component for issuing digital credentials, significantly lowering the barrier for government agencies to issue digital versions of paper credentials.[12]

The most ambitious of all is India Stack. The Indian government describes the India Stack components as allowing 'governments, businesses, start-ups, and developers to utilise a unique digital infrastructure to solve India's hard problems towards presence-less, paperless, and cashless service delivery'. India Stack includes components for digital credentials, identity and payments. Aadhaar, the digital identity component, is used by a billion users to access government services.

Common components like these make it simpler and cheaper to build public services, meaning that the teams that are building services can focus on things that are unique to their domains.

The GOV.UK Design System saves the United Kingdom around £17 million a year because the work has been done once.[13] Four years after its launch, GOV.UK Notify was saving £35 million a year, and not just because it removed duplication: GOV.UK Notify made it easier for services to proactively communicate with users, so fewer people phoned call centres.[14] Together, the GOV.UK components save every team that uses them weeks or months in development time.[15]

In the United States, cloud.gov, login.gov and the US Web Design System have also helped teams to build services faster.[16] Each project that uses the design system saves around $100,000.[17] In Argentina, a digital replacement for physical driving licences was developed from scratch in sixty-five days, in part because it made use of existing components.[18]

Time and money are only part of the picture. Technology is *path dependent*: what has gone before determines what can be done next. Because common components make it faster to build services, they can help governments to respond to the unforeseen. During Covid-19, governments had to create brand new services for things such as testing and vaccination booking, and they needed to do so at speed and under immense pressure for the civil servants involved.

The United Kingdom created 69 new Covid-19 related services by May 2020, many of them using the GOV.UK common components. By November 2020 there were at least 155 services related to the response to the pandemic that used GOV.UK Notify alone. These included notifications about test results and contact tracing, and services supporting people who were vulnerable or shielding. Users included central government departments, healthcare providers, and local and devolved governments.[19]

The Canadian federal government used the code that the UK government had open-sourced to create their own (bilingual) version of GOV.UK's Notify.[20] They used 'GC Notify' to send messages to the public about Covid-19.

India was able to rapidly create a system called CoWin for booking vaccinations. It used Aadhaar to match people's health records, and vaccination certificates were created as digital credentials that people could store in their DigiLocker account.

Governments around the world also had to rapidly change existing services such as social security and tax. There was a stark contrast between those that had mastery over their digital systems and those that did not. Many discovered how cruel technological path dependency can be.

In April 2020 the New Jersey state governor went on TV to appeal for COBOL programmers when the state's welfare systems couldn't handle the demand created by Covid-19. COBOL is a fifty-year-old language, mostly used on old mainframe systems. As another state official put it:

> Literally, we have systems that are 40 years-plus old, and there'll be lots of postmortems. And one of them on our list will be how did we get here where we literally needed COBOL programmers?[21]

Volunteers did ultimately come forward, but the lack of trained staff was only part of the problem. New Jersey's system was a complex monolith that was difficult to change without unintended consequences.[22] It also had a fixed number of mainframe computers, which meant that the state found it was unable to scale to meet demand from the public.*

By comparison, when the wave hit the United Kingdom's Universal Credit system, the dedicated digital team operating the service *was* able to respond. When there were complaints from members of the public who were unable to verify their identity online, the team could add another government identity verification component.[23] Unlike in New Jersey, scaling was possible because AWS was used to host the service.[24]

The United Kingdom may have been able to respond to the needs of people claiming welfare, but it had to rely on local government to distribute grants to business owners. It found it had no central way of sending money to businesses or communicating with them directly to understand if they met the eligibility criteria. The resulting fraud bill was described as 'eye-watering' by a parliamentary report.[25]

When payments were sent to UK business owners and people receiving Universal Credit, government payments took three days to clear because of the cost of sending instant payments over the country's banking systems. The US federal government had no way to send direct payments at all and had to resort to issuing paper cheques.[26] India, by contrast, was able to use the identity and payment components of India Stack to send $1.5 billion into the bank accounts of 30 million people digitally, with little fraud and a distribution cost close to zero.[27]

In the 1960s and 1970s the British government had hoped to rebuild a post-war sense of lost power by exporting home-grown mainframe computers of the type installed at the DVLC,

---

* In 2023 New Jersey received additional funding from the US federal government to start the move to a more modern system.

with marketing targeted at governments in Britain's former empire. The strategy failed because, as historian Mar Hicks has documented, not only were the systems poor compared with competitors from the United States, but the whole thing was bundled up in the language of and structures of colonialism.[28] It was, however, prescient about the soft power that government technology might create.

More governments are now collaborating on common components. The reuse of Notify by the Canadian government was essentially ad hoc, but India recently announced it is opening India Stack and CoWin for other countries to reuse. India's digital infrastructure has become part of the state's efforts to project its technological and economic power. Estonia has followed a similar approach with its data infrastructure.

International development funders are increasingly backing common component projects under the banners of 'Digital Public Goods' and 'Digital Public Infrastructure'. The G20 has backed the idea too.[29] MOSIP – an open-source component for digital identity similar in concept to Aadhaar – has been trialled in many countries, including Sierra Leone, Burkina Faso and Madagascar.[30] The code-sharing platform GitHub has funded community managers for several projects, including OpenFisca, an open-source 'rules as code' component.[31]

Common components are slowly being seen for what they are: core elements of a modern state that can make the public sector and society more resilient. They represent a reorganization of the work of government around new pieces of common digital infrastructure that can support many services.

PATTERNS

1. Services are made from components
2. Components support multiple services
3. Components meet the needs of teams operating services

Services are made from components

Modern digital services are made up of components, with each component responsible for a discrete task, such as looking up an address or sending a notification. A good component does one thing well. Together, they make up much of the functionality of a digital service.

The diagram shows a service for applying for benefits. It has a 'sign-in' component for users to authenticate and start their claim. The 'task list forms' and 'support chat' are used, so users can enter personal information and ask for help if they get stuck. The 'notifications' component is used to tell users about new tasks, such as booking an appointment with their assigned case worker. Finally, the 'print letter' component is used to send a letter confirming eligibility.

This sort of structure means that components can be swapped or added over time without replacing the whole service. If the benefits service wanted to replace the physical eligibility letter with a digital credential, they could add a component to do this without changing the rest of the service.

Designing digital services around components makes them flexible and able to react to changes in policy or circumstances.

Components support multiple services

If you were to look at the make-up of different public services, you'd find that they often have very similar components. Numerous services need to look up an address, take a payment or send a message.

Components can be built and operated once but used by multiple services. This means that services don't need to operate their own address lookup, payment or messaging systems. They can focus on the things that are unique to them.

The diagram shows three different services: 'Become a foster carer', 'Make a power of attorney' and 'Register land'. Each uses

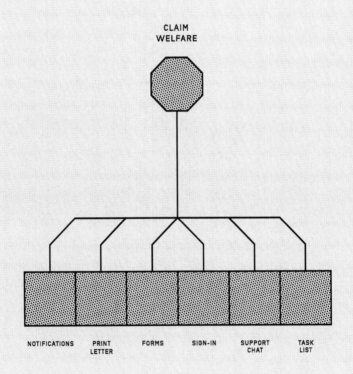

CLAIM
WELFARE

NOTIFICATIONS   PRINT
LETTER   FORMS   SIGN-IN   SUPPORT
CHAT   TASK
LIST

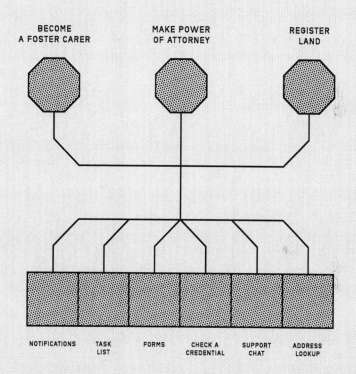

BECOME
A FOSTER CARER

MAKE POWER
OF ATTORNEY

REGISTER
LAND

NOTIFICATIONS

TASK
LIST

FORMS

CHECK A
CREDENTIAL

SUPPORT
CHAT

ADDRESS
LOOKUP

the same common components to provide part of the service: 'notifications', 'task list', 'forms', 'check a credential', 'support chat' and 'address lookup'.

Just because the components are the same does not mean that the services use them in the same way.

In the foster carer service, the 'check a credential' component lets prospective carers share their criminal record check. While the 'task list' and 'notifications' are used to tell applicants about next steps, such as booking an appointment with their assigned social worker.

In the land registration service, 'check a credential' is used by lawyers to verify their professional status and the 'task list' helps them administer a case.

Operating components in this way means the cost of individual services gets lower because service teams can focus on what's unique to their service. Furthermore, security and design problems with components can be addressed once by a skilled team whose focus is on building the best possible component rather than designing a public-facing service.

A single common component may support services across the public sector, potentially at different levels of government. For example, a component operated by central government might support services in local government and healthcare. As such, these components create public value well beyond the institution that operates them.

Components meet the needs of teams operating services

Common components make it faster, cheaper and easier to create services. That's because the teams building services are freed up to focus on the things that are unique to that particular service. They still create the bespoke parts of the service, as well as the managing rules and the data for the service, but they don't have to operate the components.

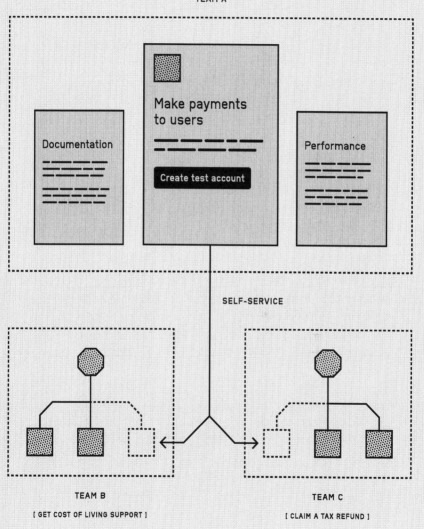

TEAM A

SELF-SERVICE

TEAM B

[GET COST OF LIVING SUPPORT]

TEAM C

[CLAIM A TAX REFUND]

The teams building and operating services are the primary users of a component, not the public, who experience them as part of a wider service. This means components need to be designed to meet the needs of those teams, and they must be findable and usable without direct support. Onboarding processes that require meetings and components that need customization before they can be used represent significant friction for digital teams who are working at pace.

The diagram shows a 'Make payments to users' component. It is being used by two different services: 'Get cost of living support' and 'Claim a tax refund'. The payment component has been designed for 'self-service'. The work of integrating with the payment component is done by the teams operating the services rather than the team operating the component.

Components that are designed for self-service explain to their users how they work and how to use them. Self-service lets service teams check whether or not a component meets their needs. This means that these components need a public website that includes technical documentation, sample code, and security and procurement information. They may also provide sandbox environments (a place where software engineers can test code without affecting the live system) that emulate the 'try before you buy' approach of commercial Software as a Service providers. As Pete Herlihy, product manager at the launch of the GOV.UK Notify component, put it: 'Jeff Bezos doesn't come round to your house for a chat when you're thinking about buying something from Amazon.'[32]

Components also demand an operating and funding model that allows them to be used without protracted procurement processes. This means that the needs of procurement and finance officers need to be met, as well as the needs of software engineers.

Build components that solve problems once

Common components are an achievable aspiration that can transform public services, but experience shows that they are unlikely to emerge organically from government agencies. The institutional incentives are not strong enough.*

Creating successful common components requires identifying potential components and then building an effective team around those components, supported by an appropriate funding mechanism. Potential common components can be identified by

1. conducting interviews and surveys with service delivery teams;[33]

2. using 'Wardley mapping' to identify duplication;

3. collecting data from governance processes, such as service standard assessments, to understand common needs; and

4. reviewing code repositories to identify teams that are working on the same problem.

In addition to identifying duplication, there need to be clear routes for turning 'point solutions' (problems solved within individual services) into common components, for example, through supporting them with additional staff and funding. Internally, both Amazon and Google have processes for this. They can create and fund a new team off the back of something that is happening organically.[34] There are examples of this happening in government too. Login.gov started life as a larger project called MyUSA that aimed to provide a single account and task list for interacting with the federal government. The authentication part of the project was developed

---

* No common components emerged from the UK Government Digital Service's 'exemplar programme', which saw teams set up across government to redesign digital services.

into login.gov by 18F (a digital services agency within the US General Services Administration) and the United States Digital Service.[35]

Infrastructure without organizational change is destined to fail. Components need teams with clear mandates and permission to start small. They also require an institutional home that is not too closely bound to a particular policy department.

The creators of Aadhaar talked about creating a start-up within government. They started small with a cross-disciplinary team, and then they grew. Aadhaar started life in the planning ministry: outside a policy-based silo. It was placed there to avoid 'turf wars between government ministries' and because 'projects developed in one ministry may not be adopted by others'.[36] Eventually, it became its own organization (see chapter 10).

Components need funding for the long term *as infrastructure*. Traditional funding models such as cost recovery and project-based funding don't suit them. Cost recovery excludes smaller-volume users, and services will be less likely to use a component if they think it might disappear at the end of a funding cycle.

Component teams must work together with finance teams to develop long-term financing that takes account of the uncertainty that exists in the early stages of a component being developed. The government of British Columbia's J-P Fournier has translated these characteristics into 'accountant-speak' as follows:

> Digital funding should match the rhythms of digital delivery teams. ... Software is an asset class that depreciates fast and has high maintenance requirements, so you need to fund it appropriately.[37]

Component teams working with finance professionals can also build an understanding of how long it might take to reach scale and take account of the full societal value that might be created.[38]

Main points

1. Many governments have been digitizing paper processes within the boundaries of individual agencies.

2. This has created 'technical debt' in the form of duplication, as well as inconsistent user experiences.

3. There is a latent cost of services that have never been digitized because the cost is too high.

4. Complex systems are hard to change.

5. Digital systems are made of components.

6. Some governments are creating common components that support multiple services.

7. Components do one thing well.

8. Components create resilience by making it faster and simpler to create services.

9. Some governments are exporting their common components.

10. Components are designed to meet the needs of teams' delivery services.

11. Components should be operated by dedicated teams and appropriate funding.

12. Designing for self-service means that you don't need a meeting to start using a component.

13. Potential common components can be identified using surveys or Wardley mapping, by collecting data from governance processes, and from code reviews.

14. There should be a route for 'point solutions' to be turned into components.

Examples: anatomyofpublicservices.com/examples/components

CHAPTER 6

Data as infrastructure

In 2013 the UK sold its postal delivery infrastructure and accidentally sold part of the country's data infrastructure along with it.

The British Post Office had been sliced and diced over the preceding decades. The telecom network was split off in the 1980s, and British Telecom and Girobank were sold in the 1990s. In 2013 it was the turn of Royal Mail and its network of sorting offices, its postal workers, its fleet of vehicles and 'the PAF', or Postcode Address File.

The PAF is the canonical list of the country's postal addresses and postcodes.[1] It's arguably the most familiar dataset in the United Kingdom. More people know their postcode than their PIN or mobile number.*

The PAF is used for everything from ordering a pizza to telling people where they can vote, from mapping apps to reporting potholes in roads and applying for benefits. It's as much a part of the United Kingdom's national infrastructure as the road network.

The data could have been opened up as a public good, but privatization meant that the public sector would have to rely on a licensing agreement, which would (hopefully) be renewed every few years. Most companies and charities would have to pay to use it.

---

* The format of the postcode was designed by a team of psychologists in the 1950s to be memorable.

The PAF had been bundled into the sale of Royal Mail in an attempt to boost the share price, but judging by a parliamentary review a few years later, it's not clear ministers realized what they had sold. Criticizing the 'short-term gain' of the decision, the review said:

> The sale of the PAF with the Royal Mail was a mistake. The Government must never make a similar mistake. Public access to public sector data must never be sold or given away again.[2]

Bureaucracies run on lists. A search on legislation.gov.uk for the term 'a register of' reveals official lists of voters, companies, medicines, firearms, land, landlords, common land, adopted children, trains, burials, rocket launches, establishments that keep laying hens, police complaints systems, cows, childminders, village greens, green gas and insolvencies among the seventy-eight pages of results.[3] But not all lists are created equal. Some, like the PAF, support critical services. Lists of land ownership, voters, charities and so on are the 'foundational data' that a state needs to operate.

Foundational data creates expectations and rights. An expectation that a parcel will be delivered; the fact that someone has the right to vote or that a donation went to a genuine charity. This type of data places responsibilities on the state, to maintain records or to regulate. For example, this is what Massachusetts law says about company registration:

> One or more persons may act as the incorporator or incorporators of a corporation by signing articles of organization and delivering them to the Secretary of State for filing.[4]

It gives the Secretary of State a clear duty to place information about new companies on record. And this is how a Welsh law creates an organization and tasks it with maintaining a list of landlords:

A licensing authority must establish and maintain a register for its area.[5]

Many official lists have a long history. Her Majesty's Land Registry was established in 1862 to register the ownership of land in England and Wales (by 2024 it had registered 88% of the land mass).[6] But even newer lists, or 'registers' as they are often known, tend to follow the patterns of paper-based organizations. This is because it is costly to duplicate, reuse, recombine and search paper documents, so organizations were set up with the sole job of collecting and maintaining records.

In the case of HM Land Registry, the law limits the Chief Land Registrar to creating 'forms' where it relates to the business of registration:

The registrar may prepare and publish such forms and directions as he considers necessary or desirable for facilitating the conduct of the business of registration under this Act.

The registrar can choose to publish other information if they perceive a public interest (and, presumably, if they *choose* to prioritize it), but it's not their core mission:

The registrar may publish information about land in England and Wales if it appears to him to be information in which there is legitimate public interest.[7]

In comparison with paper records, data *can* be duplicated, reused, remixed and searched easily. But organizational contexts like this mean that foundational data tends to remain tightly coupled to the services that collect it. Land registration data is therefore collected for registering land. Earnings data is collected for calculating income tax. Property data is collected for calculating property taxes (but may not link back to address data collected for the purpose of putting letters through letterboxes). Company information is collected for the purpose of registering companies (but may not link to data collected for calculating company taxes).

In the public sector, the solution is normally sought in the *sharing* of data. 'Data sharing' is one of those magic phrases in meetings that triggers a guaranteed response: most will agree it's a good thing but there is a knowing understanding that it's hard and unlikely to happen. Some will be opposed to it on principle, fearful that the consequences of sharing any data would be misunderstood. Institutional folklore will be retold (every department carries an 'if only' list down the years), with stories of grumpy lawyers or unhelpful IT managers. There will be rumours of new systems and teams in other departments to be followed up.

What people *mostly* mean when they talk about data sharing is that they have a problem and they think data from somewhere else can fix it. 'If only the tax office could tell us when a bill has been paid, we wouldn't need form TRP15 anymore, we could update our system automatically!' And so on.

Rarely though is there a common understanding of what a good *implementation* of data sharing might look like. The result of this fuzziness is that the term replicates wildly. In 2024 a search on the GOV.UK website for 'data sharing' returns more than 5,000 policy documents, news articles and research papers. But 'data sharing' is a euphemism. In reality, it's closer to copying and pasting. Data gets duplicated and is immediately out of sync and out of date. Actually, it's more like faxing. The copy is a lower-resolution version of the original. It is partial.

To give an example of this, the United Kingdom's Universal Credit welfare system uses earnings data from the country's tax office, HMRC. Their data sharing agreement shows that Universal Credit receives a partial copy of the data from HMRC based on a set of National Insurance numbers. These get sent to HMRC once a day, who then send the data back four times a day (Monday to Friday). When the numbers don't match, or the data is out of sync, users can get stuck in limbo. Universal Credit and HMRC then have to spend time and effort investigating.[8]

'Data sharing' also suggests that the thing being shared remains the same. But you wouldn't ask someone to share their bicycle with you, then ask if you can add another wheel! Because official data tends to be highly bound to a particular service or a particular organization, data collected for one purpose is rarely directly usable for another purpose.

What Universal Credit considers income, for the purposes of paying benefits, is not the same as what HMRC considers income for the purposes of charging income tax, the result of which is that Universal Credit has to add to the data it receives to include other sources of income. Similarly, the list of other countries a state legally recognizes is not the same as the list of countries whose identity documents it needs to process for visa applications. The list of registered companies is not necessarily the same as the list of organizations who have to pay company taxes.

'Sharing' also suggests a one-off event. That might be the case when asking someone to share a spreadsheet used to create a chart in a presentation, but it's not for data that supports public services. Requesting data to be 'shared' for that sort of use is asking another organization to use its limited capacity to meet the business needs of the requester, and to do so *into the future*. It is, essentially, asking another organization to reprioritize work, change its systems and process, hire staff and dedicate time to meetings.

The political importance of Universal Credit meant that the institutional inertia to share data was overcome, but in general, data remains fractured and fragmented across the public sector. The goto fix for this compartmentalization is to try to improve the process of cross-government data sharing on its own terms, such as coming up with a better process for creating data sharing agreements, or implementing open data 'portals'. But strategies like these attempt to increase the consumption of data without changing its production.

Mostly, these attempts to improve data sharing are examples of institutional procrastination. Or rather, 'bikeshedding'. In software engineering, 'bikeshedding' means spending a disproportionate amount of time on the trivial parts of a project. The term was popularized in geek culture by a weeks-long argument on an email list of the open-source FreeBSD project about whole numbers versus fractions. It is a reference to Parkinson's Law of Triviality, coined by Cyril Northcote Parkinson, a British naval historian and public administration academic.[9] Parkinson imagined a finance committee whose job it was to approve the designs for a nuclear power station. Fazed by the complexity and expense of the reactor, they assume that the people working on the reactor *must* understand it so they skip over the details. They spend the bulk of their time discussing item 10 on the agenda: the staff bike shed. Committee members want to make a contribution to show they are doing their jobs. The bike shed is familiar and the problem is tractable.

Making it a bit easier to agree a data sharing process does little to remove the cost of implementing and operating it into the future. Making it easier to publish data on a 'portal' does nothing to bind the source of the data into understanding the needs of those who use it or managing data as a common good.

Focusing on better data sharing in government is the same as talking about what colour the bike shed should be. The elephant in the room is that systems don't talk to each other and the data is all at angles. Just as digital infrastructure without organizational change stores up problems of duplication (see chapter 5), so organizational change without new infrastructure creates inertia.

People are able to imagine a spreadsheet being sent between organizations. They are able to imagine a change to a legal agreement to share information. They are able to imagine meetings in which they agree the data fields that will be used and how often the updates will happen. The result is that those are the things that people talk about and organize their ideas around.

To create modern public data infrastructure is to change how data is *organized* and how it is *accessed*. It must be organized to minimize duplication and maximize interoperability. And it must be accessible by services in a way that is predictable and involves minimal friction.

Luckily, organizing data so that it has less duplication and is simpler to join with other datasets is an idea nearly as old as databases. In 1970 Edgar F. Codd, a British computer scientist at IBM, published a paper called 'A relational model of data for large shared data banks'. The paper introduced a new type of database: the 'relational' database.[10]

Databases had been around since the 1960s, but data was stored in rigid hierarchies or networks. So a database of government departments, policies and projects would be represented as a tree. Projects would be stored as branches of policies, and projects as branches of policies. In these 'navigational databases', programmers needed to understand the links between the data before they could 'navigate' the data. They often had to write specialized programs that ended up tightly bound to the structure of the data. If the data structure changed, the program broke. Data was also duplicated. If you wanted to represent a project that covered two policy areas, then it had to be stored twice: once under each policy area.

Codd argued that data should be stored independently of how it was used.[11] In a relational database, the data would be 'normalized', meaning that it was stored in its simplest form in tables. Programmers needed to know only the structure of those tables. So, in the example of government departments, policies and projects, there would be a table for each. Each table would have a 'primary key' used to identify each row in the table. Those IDs could be added to other tables to describe the relationships between them. So, a table containing project IDs *and* policy IDs can represent projects that span multiple policy areas.

IBM was not very happy about Codd's work. At that time, IBM made money selling hierarchical databases, not relational ones, so when the paper started receiving attention Codd was accused of undermining the company's 'strategic' products. The idea of relational database nearly died, but Codd persisted.[12] IBM would eventually launch its own relational database product (in part because of the US government funding competing work, legitimizing Codd's work). The relational model would go on to become the de facto standard for data storage, used in everything from buying online shopping to renewing a driving licence. It created an industry worth billions of dollars, and Codd received the Turing Award in 1981 for one of the great technical achievements of the twentieth century.[13]

Countries, such as the United Kingdom, that have lots of legacy technology have their data organized much like the navigational database model that Codd was aiming to replace. Before the advent of relational databases, programmers had to understand the detail of how data was stored and write bespoke, brittle software to access it. Today, a team building a new public service needs to divine what data exists, in which departments, what form it is stored in, and how to match it to the data they already have. They then need to create both an ad hoc data sharing agreement and the software to keep the data in sync.

Some countries, most notably Estonia and India, are organizing their foundational data with more intent. Less weighed down by legacy technology, they have positioned data as a common resource for wider society.

In Estonia data is stored in databases that are distributed across government departments, but a law prohibits them from holding data that is already held in other databases:

> Establishment of separate databases for the collection of the same data is prohibited.[14]

Data is stored by domain (motoring, health, tax) and then linked using standard identifiers. Essentially, the data is normalized.

The result is that when Estonians renew their driving licence – a process that requires a health declaration – their application can retrieve data from their patient records.[15] Future mothers and fathers can see how much parental benefit they will receive using a calculator that is pre-populated with data from the Estonian Health Insurance Fund and other government databases. Interest from bank accounts can be sent from online banking directly to the tax authority, so tax returns can be pre-filled.[16] Even graveyard databases link back to the population register, so relatives can request maintenance and memorial services.[17]

The custodians of each database must publish a list of the types of data they hold in a central register. This is partly a transparency measure (see chapter 9), but it also acts to help public servants meet the once-only principle (see chapter 1) of not collecting the same information twice. There are also processes for auditing to ensure the rules are being followed.

India, at a very different scale of population (1.35 million for Estonia versus 1.42 billion for India), is following a similar model. The Unique Land Parcel Identification Number will be a unique identifier for every land parcel in the country, replacing separate identifiers used in different Indian states. The fourteen-digit alphanumeric ID is generated from the latitude and longitude of the vertices of land parcels.[18] The aim is for the number to be used across government and wider society, joining up data from the courts, land registration and planning, and giving farmers access to finance.

India is creating standard identifiers for other domains too. In health, a patient's ABHA number brings together data from across the health system and combines it into a 'longitudinal health record', linking together different public health programmes and insurance schemes.[19] The nation's health facilities are being recorded in a single Health Facility Registry too, linking them to people's health records and making it simpler for hospitals, pharmacies and clinics to renew their licences.[20]

In education, the Automated Permanent Academic Account Registry aims to make it easier for students to transfer between schools and access their records throughout their life.[21]

Under a proposal by India Post, postal addresses may soon be replaced by a unique identifier too. Digital Address Codes generated from geospatial coordinates would aim to enable better targeting of benefit payments and e-commerce.[22]

Estonia and India have shown that, far from being an aspiration, 'data sharing' is an anti-pattern.* What is different about the Estonian and Indian approaches is not the creation of identifiers – after all, the United States has Social Security numbers and the United Kingdom has National Insurance numbers – it's that data across the public sector is being organized around those identifiers and the foundational data they represent.

It's not enough for there to be standard identifiers and tidy databases, however: how the data is accessed is important too. In both India and Estonia, data can be accessed at source from a wide range of public sector and private sector systems. This is made possible by 'data exchanges'. These exchanges make data findable and accessible via APIs. They also enforce access, security and transparency.

In addition to the ABHA number, India has created a dedicated agency – the Ayushman Bharat Digital Mission – to create the infrastructure needed to access data across the health system. The Unified Health Interface will eventually allow services to access data about things such as appointments and test results using the internationally recognized FHIR (Fast Healthcare Interoperability Resources) protocol.[23] There are similar efforts underway with agricultural data and urban data.[24]

In Estonia, X-Road is a data exchange that all government agencies, and some non-government organizations such as banks, use to access data. This, in combination with unique

---

* In software engineering, an anti-pattern is a commonly used solution that has more bad consequences than good ones.

identifiers and the no-duplication rule, is what makes it possible to prefill tax returns and benefit calculators.

Because neighbouring countries have also implemented X-Road,* data can be accessed across borders: Finnish prescriptions are recognized in Estonia, and vice versa, for example.[25] Finland, Estonia and other Northern European countries are also beginning to support cross-border invoicing and logistics reporting in real time.[26]

Organizing data in the way first described by Codd in the 1970s (stable identifiers, less duplication, stored in its simplest form) and building dedicated infrastructure to access that data creates public value beyond the walls of the organization responsible for that data. Estonia claims that X-Road saves the public 1,345 years of working time every year.[27] India is hoping that the Unique Land Parcel Identification Number will make a dent in the 1.3% of GDP that is lost to stalled projects and land disputes clogging up the civil courts.

But data infrastructure is not just about efficiency. As Estonia's cross-border services show, it can redefine the spaces in which public services operate. The boldest attempt at this is India's Open Network for Digital Commerce (ONDC): a non-profit company, tasked by the Indian government with creating an open network for the exchange of e-commerce transactions. Through data standards and APIs, it connects customers, sellers and delivery providers.

Initially, it launched as a small-scale beta version in Bangalore that linked the public with grocery sellers. As of February 2024 it had handled 7.5 million transactions for everything from metro tickets, ride hailing, show tickets, groceries and personal care.[28] It has the potential to open up new policy interventions, too. In the midst of a food price spike in 2023, the government was able to use ONDC to sell cut-price tomatoes over the network.

---

* Estonia published the code as open-source.

In time, the ONDC could create a public interest rival to e-commerce platforms such as Amazon. While the United Kingdom may have 'accidentally' sold off its data infrastructure along with its delivery infrastructure, India is attempting the same manoeuvre in reverse: creating a national delivery infra-structure off the back of its data infrastructure.

PATTERNS

1. Data supports multiple services
2. Data access (not data sharing)
3. Data infrastructure works across boundaries

Data supports multiple services

The value of data is greater than the use it is collected for. Managed in the right way, one dataset can support services across society.

The diagram shows three public-facing services:

1. a city government service for paying a city tourism tax;

2. a commercial service for renting a holiday home; and

3. a central government service for paying tax on the sale of a property.

All three rely on the same data: a register of landlords, a register of property ownership and a list of addresses.

The tourism tax is payable when people visit the city and stay in hotels and holiday properties. The service uses landlord and property ownership data when enrolling the managers of hotels and holiday properties onto the scheme. Then it uses the *address* data for tourists to verify the address where they are staying.

The holiday rental service uses the address data for guests to find a property to stay in, while property data is used when owners join the service. If a property is rented out for long periods, the owner must register as a landlord. The service helps owners through this process and sends the data to the register of landlords.

The property tax service uses the register of landlords and property ownership to calculate the correct amount of tax payable. (In this example, landlords who own multiple properties pay a higher rate of property sale tax.) The address data is used to look up the property at the start of the payment process.

Supporting multiple uses of data like this is only really practical if the data is stored and managed properly. This means using common identifiers and storing the data once, in its simplest form.

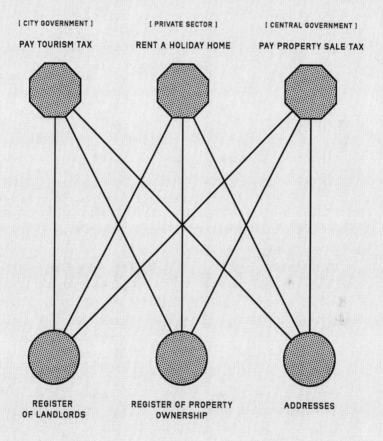

[CITY GOVERNMENT]   [PRIVATE SECTOR]   [CENTRAL GOVERNMENT]

PAY TOURISM TAX   RENT A HOLIDAY HOME   PAY PROPERTY SALE TAX

REGISTER
OF LANDLORDS   REGISTER OF PROPERTY
OWNERSHIP   ADDRESSES

Data access (not data sharing)

Historically, supporting multiple uses of data has been done through what is euphemistically called data sharing. But 'sharing' data is, essentially, copying and pasting from one database to another. The data gets duplicated, creating redundancy. It can easily get out of sync, which creates problems for users and the teams operating services. Furthermore, because data is tightly bound to the purpose it was collected for, it often has to be transformed before it can be used.

A better approach is for data to be managed as a common resource and *accessed* via an API. The diagram compares the *data access* approach with *data sharing*. It shows three services using data about people's earnings: 'Pay income tax', 'Claim welfare' and 'Apply for free childcare'.

In the data sharing example, each night a copy is made of a subset of data and sent to the social security department and the education department, creating three separate copies, all of which have to be secured and maintained.

The earnings data is tightly bound to the income tax service, and it is not therefore 100% suited to use in the other two services. As such, as well as the data having to be duplicated, it has to be transformed and augmented with additional data fields. Some data is also removed to avoid sharing sensitive information.

In the *data access* example, by comparison, rather than each service maintaining its own full or partial copy of the data, there is a single, common database that is optimized for interoperability. Rather than sending copies back and forth, data is accessed using an API. Access control rules ensure services can only access the information they need.

Managing data as a common resource in this way requires a custodian who can prioritize the needs of multiple services, as well as being accountable for the quality and integrity of the data. Custodians need to understand the needs of their current

Data sharing (legacy)

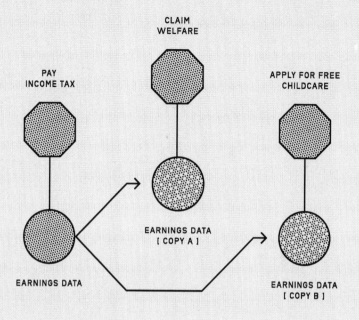

Data access

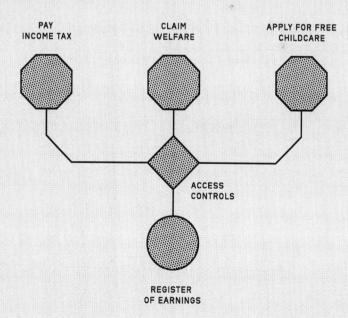

and potential users. They need to ensure their APIs are designed for self-service, so that users can find them and understand how to use them unaided (see chapter 5).

Data infrastructure works across boundaries

Data infrastructure can make services work across jurisdictional and organizational boundaries. Those boundaries could be between two nation states, but equally they might be between two health authorities, each running its own network of hospitals.

The diagram shows a service designed to help companies export their goods. It does two things. First, it helps exporters comply with rules in their own country, handling tax payments and issuing credentials, so that their goods are ready to leave the country. And second, it helps them comply with rules and taxes in the country where the goods will be *imported* to. It does this by transferring data about the goods along with the necessary export certificates to the importing country. Once the goods have been cleared, the service can request any additional certificates from the importing country.

Services like this allow one state to reach into the administrative space of another. They meet the needs of users, beyond borders. They also have the potential to create both a real-time experience for users and real-time data for the governments operating the data infrastructure.

Interactions like these are made possible when there are common standards for exchanging data and the underlying rules are exposed as APIs. But it is not just about technology. There needs to be mutual recognition of data infrastructure and standards, and (most importantly) there needs to be political alignment between different jurisdictions.

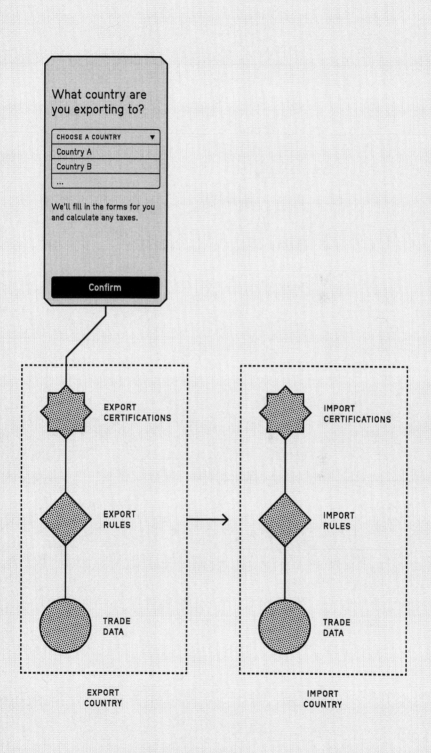

Decouple data from services

To maximize the value created, the one-to-one relationship between service provision and foundational data needs to be broken. However, unlike components that can provide value to a service from the day they are launched, pulling off the same trick for data is harder. There is a critical mass at which data infrastructure becomes useful.

A component for sending messages is, from the perspective of a service using it, just as useful on day 1 as it is on day 1,000. It does one thing well, and it keeps doing it. With data, it becomes more usable the more it is used and the greater the number of other datasets that link to it.

For example, a register of patients that only links to bookings at one hospital is less usable than one that links to bookings at every hospital. If it also links to blood test results, it becomes more usable still. This pattern shows in the growth of Estonia's data ecosystem. Only when fifty datasets had been connected to the X-Road data exchange did the number of data access queries start to grow exponentially.[29]

Because the time until a threshold like this is met might be long, work to break the one-to-one link between services and data needs to be funded for the long term. It is unlikely to result in short-term cost savings. Estonia and India do, however, show that it is possible to start small with real use-cases, so the initial funding might not need to be huge.

Organizing the effort to make data into a common resource around a clear set of user needs and administrative burdens can mitigate the risk of repeating the failings of the open data projects of the early 2000s. These tended to focus on openness for the sake of openness, and they failed to create the change they hoped for because they didn't understand the needs of the consumers of data and found it hard to find a place in the day-to-day work of the custodians of data.

The aim of planning.data.gov.uk, by comparison, is to enable data-driven decision making in the property sector, so that builders know where to build houses and urban planners can do better planning.[30] It brings together data from hundreds of local governments in standard formats, including everything from flood-risk data to building-design codes.[31] In 2020 the UK government also published its Unique Property Reference Number under an open licence, meaning it can be used in any database without paying licence fees.

One final option is to follow the Estonian model and legislate to change how data is managed in the public sector. When Estonia passed the law that databases must not duplicate data that the government already held, it did something else at the same time: it defined what a database was. The requirement not to duplicate along with requirements for transparency were applied to any databases that met that definition.

This sort of redefinition is not that uncommon. Some countries even have a legislative process, known as an 'interpretation act', for clarifying and defining concepts that have been mentioned in prior legislation.[32] Interpretation acts do things like stating that the word 'distance' means 'measured in a straight line on a horizontal plane' (presumably to stop people claiming tax relief on a journey measured as an ellipse).[33]

Could a 'data interpretation act' redefine the word 'register' or 'database' as something with a custodian, an API and no duplicated data? For jurisdictions without interpretation acts, a change to the general record keeping obligations of an official, such as a Secretary of State or mayor, might be required, but the idea is still the same.

Sometimes people argue for maintaining the messy status quo of public sector data as a protection against government overreach. But that's like arguing against fixing potholes as a road safety measure instead of arguing for the speed limit to be reduced. There are definitely situations when data should remain tightly bound to a service and when it should be very, *very* difficult to join data together. But these decisions should be the result of informed policy choices. They are too important to rely on poor housekeeping as a safeguard.

Main points

1. Governments run on lists.

2. An authoritative list required for operating a modern state is termed 'foundational data'.

3. Foundational data creates expectations, rights and responsibilities.

4. Foundational data supports multiple services.

5. Data tends to be tightly bound to the services that collect it.

6. Data sharing duplicates data and places ongoing expectations and costs on the holder of the data.

7. Data access means using data from a common database via an API.

8. Normalization of data reduces duplication and creates unique identifiers.

9. Data needs custodians with the right mandate.

10. With common standards, data infrastructure can work across borders.

11. To create data infrastructure is to change how data is organized and how it is accessed.

Examples: anatomyofpublicservices.com/examples/data

CHAPTER 7

Empathy augmented

The social scientist and psychologist Sherry Turkle has researched people's relationship with technology from the days of the 1970s hobbyist hackers through to today's blanket use of smartphones. She puts what she calls a 'crisis of empathy' – a 40% drop in markers for empathy in US college students, mostly in the period to 2015 – down to the use of digital.[1] Partly, she argues, this is the nature of phones as devices that 'block' face-to-face interaction, but it is also a result of the medium of always-on text-based communication:

> Human relationships are rich, messy, and demanding. When we clean them up with technology, we move from conversation to the efficiencies of mere connection. I fear we forget the difference.[2]

Just as our devices act as blocks to conversation, could digital interfaces block our relationship with the state?

Empathy is the understanding of the feelings, thoughts and experiences of others. It relies on 'mentalization': the human ability to build and maintain a working model of others' minds. We do this by inferring and interpreting people's behaviours and social cues, based on what we know of their situation. Although, regardless of the available cues and context, people's ability to mentalize varies.[3]

However, today, digital services tend to be individual experiences: an interaction between 'the user' and the surface of the service. The people behind the service and the decisions they make are hidden – put at a distance. They don't do small talk. Furthermore, because of digital systems' ability to abstract away networks (see chapter 2), where there is a direct relationship with officials, it is likely that each interaction will be with a different person, with limited access to the previous context.

For most people, getting a driving licence or a passport will never have included much by the way of conversation. But for public services that are inherently messy, rich and demanding – such as healthcare, social security or public housing – does putting the relationship behind a form or filtering it through a chat interface dilute the relationship that people say they want with public services, in the pursuit of efficiencies?

Even for more transactional services, might we have lost something in the move to digital? It's been argued that self-service checkouts in supermarkets and touch screens in take-away restaurants may increase social inequality because they reduce the chances we have to interact with other people.[4] Digital services mean less waiting in post office queues and fewer conversations with government caseworkers on the phone.

One argument is that the relationship with government doesn't matter so long as the public gets the right outcomes, but it turns out that people don't care only about outcomes – receiving a welfare payment in their bank account, say, or receiving their driving licence in the post – they care about how the process makes them *feel*. They ask: did it make a good-faith effort to help? Are decisions objective and even-handed? Are interactions consistent? Are burdens fairly distributed between the public and officials? Does it feel *just*? *Fair*?[5]

Things that feel unfair, inconsistent or subjective create psychological costs for users. Psychological costs are the third type of administrative burden defined by Pamela Herd and Donald Moynihan (see chapter 1), and their research shows that fairness is something that is 'fundamentally important' to the public.[6] However, as Joe Tomlinson, the director of the Administrative Fairness Lab, points out, there is 'little evidence on what the public thinks fairness looks like', which means that services are not 'shaped around what the public thinks is procedurally fair'.[7]

In 2023 the Administrative Fairness Lab conducted research into the Universal Credit welfare system, and its findings were similar to those of Herd and Moynihan: people claiming Universal Credit prioritized the need for interactions to feel respectful, dignifying and empathetic.

Users had positive experiences when, among other things, they felt listened to and had a voice in the process. However, chat messages that were terse or blunt, and phone calls and face-to-face conversations that were perceived as dismissive, felt disrespectful. If they were punished for errors they made but observed errors by the government going unaddressed, they were left with a feeling of double standards. They also preferred having interactions with the same caseworker over time because new relationships came with a risk of being misunderstood or of advice being inconsistent. Overall, the public defined their interactions with Universal Credit in terms of a relationship and it was the quality of that relationship that mattered most, even before things like ease of use.[8]

There is an extreme example of what happens when human relationships and fairness are not considered as part of the design of a service in the form of a website that predated the launch of Universal Credit: the ill-fated Universal Jobmatch.*

In 2013 the British politician Iain Duncan-Smith said that 'looking for work should be a full-time job'. Welfare claimants having to prove they are looking for work had been a component of the British welfare system since as long ago as 1921, when the 'seeking work' test was introduced.[9] But Duncan-Smith's proposal was for a tougher approach – one enforced by a digital channel.

---

* The UK-wide launch of Universal Jobmatch predated Universal Credit.

In contrast with the paper form and face-to-face meetings that it replaced, government caseworkers could directly monitor how users were searching for and applying for jobs on the Universal Jobmatch website, and check if they were meeting the thirty-five hours per week target. In theory, use of Universal Jobmatch was optional, but a civil service union complained that managers were putting staff under pressure to tell users it was mandatory, so most users probably felt like they had no choice.[10]

A couple of things happened in response to the launch of Universal Jobmatch. First, because anyone could post a job on the website, and because people tend to waste little time in grasping opportunities to stick it to the man, people started posting spoof job adverts for things such as 'MI5 assassin'. An anonymous group of benefits campaigners also released a browser plug-in that automatically submitted random job applications on behalf of users. All a user had to do to comply with the hourly time requirement to look for work was click a button, the campaigners claimed.[11] Second, and much more harmful, there were reports of users being scammed by false employers who were requesting payments for criminal record checks.[12]

There are many ways in which Universal Jobmatch was a failure. Partly it was a failure of implementation: a service designed upfront by policymakers and delivered by an outsourced company, all in the absence of any understanding about how users might react to it.[13] But it was also a failure because it designed out empathy on both sides of the relationship between welfare claimants and government caseworkers. The whole stance of the service was adversarial.

Technology can make the world smaller, connecting people instantly, regardless of their location and allowing them to interact at any time (see chapter 2). But Universal Jobmatch showed another feature of this phenomenon: while the world becomes more connected for some (e.g. caseworkers can monitor more welfare claimants), others become marginalized.[14]

Universal Jobmatch scaled only one side of the human relationship: users were watched but they could not see that they were being watched.

Blocking empathy is not the only way that digital may change the relationship between the public and public services. Digital can also impose its own normality. As designer Dieter Rams said: 'Indifference toward people and the reality in which they live is actually the one and only cardinal sin in design.'[15] Could digital public services leave more people feeling unheard and misunderstood?

Storing data in a database and writing software code implicitly create rules about what is 'correct'. The website 'Falsehoods Programmers Believe' maintains a list of normalities that are regularly imposed on users of digital services across the public and private sectors. It includes things such as the assumption that everyone has a first name and a second name, assumptions about the size and shape of families, and assumptions about the format of addresses.[16] If a service cannot store someone's name or address in a way that person recognizes, then a service cannot talk to them properly.

Historian Mar Hicks identified an early example of this phenomenon. In the early 1960s a new computer system at the UK Ministry of Pensions and National Insurance ended the ability for trans people to have their names reflected on their National Insurance card and in correspondence.[17]

More recently, Netflix and other streaming services have found themselves having to define a 'household', which is probably a task without a correct answer.[18] Of course, the law is quite capable of creating these sorts of normalizing effects too. The UK government defines a couple differently for the purposes of taxation and welfare payments, for example. But it is easier to impose norms through digital systems than it is to change the law.

Could digital public services also be inherently less valued by the public?

Despite the advantages of digital versions of books, films and photos, people place more value on the physical versions. A 2017 study by the marketing researchers Ozgun Atasoy and Carey Morewedge found that people are willing to pay more to own physical copies.[19] They found that this is partly due to the additional control people perceive they have over physical products. They also found that physical goods generate a greater sense of 'psychological ownership': the feeling people have that something is *theirs*. People can feel it for a whole range of things, from their possessions to their local park or a company they work for.

In the context of public services, might people feel less ownership and control over digital forms and digital credentials than they do with paper forms and plastic cards?

Certainly, paper forms can be 'completed' in any order. Incomplete forms are a way to ask for help from friends, family or government officials. A copy can be saved for the future or as a way of 'comparing notes' with others. Digital forms tend to be completed in order and then submitted once complete. When a form is submitted or a digital account is closed, the information is gone.[20]

Physical identity documents are closely linked to our sense of personal identity. They can be hidden away in files and at the back of cupboards, kept alongside related documents. But digital credentials live on a device and can disappear when remotely revoked.

There is hope, however, that digital services can create a good sense of psychological ownership. Atasoy and Morewedge suggested involving users in the customization of digital products and the use of 'skewmorphic design' as ways of increasing the feel of ownership and control. Skewmorphism – the approach of designing digital interfaces to look like the real-world version – fell out of fashion in the early 2010s. However, in 2024 digital versions of books, music and bank cards are the last refuge of the approach. You will still find it in how Apple, Amazon and

Google represent those things. Maybe those companies are try-ing to recreate the sense of control and ownership?

A US study of government benefits also found that simply using language that increased a sense of psychological owner-ship when talking about government benefits significantly influ-enced people's desire to make a claim online.[21]

Designers of digital public services may find they need to make more space to understand the qualities that are lost in the move to digital, and think about how to fold some of those qualities back in.

Of all these changes that digital brings, the question of what it does to human relationships is the most challenging. Is it pos-sible to 'scale the human'? To replicate the effects of empathetic human relationships at a distance in space and time, and to do so in greater numbers than is possible with face to face?

People have been trying for a long time. In 1976 the British Post Office funded research by John Short, Ederyn Williams and Bruce Christie into people's attitudes to different communica-tions mediums. The Post Office had developed 'Confravision', the first public video conferencing system aimed at businesses and government. Meetings were held in dedicated high-mod-ernist TV studios around the United Kingdom, with cameras controlled by the chair of the meeting, with a switch to turn on the encryption 'scrambler'.[22]

Comparing phone calls, video calls and letters, Short and his colleagues identified issues familiar to everybody today with video: 'The visual channel available in most video systems does not restore eye-contact as a cue; it makes things even worse'; 'The camera cannot be placed exactly in line with the picture of the eyes, so if person A thinks he is looking person B in the eye, he will appear to B to be looking elsewhere.' Other noted issues were the 'tea and biscuits problem' that arises if some people are remote and others are not, as well as the loss of 'chats at the beginning and end of the meeting, the break for lunch and the drinks in the bar'.[23]

Confravision was a flop, but the research founded 'social presence theory'. Social presence is a measure of the sense of being with others, and the sense that the relationship is truly represented. Social presence theory studies how it changes with different communications mediums.[24] As expected, in-person interactions score highest because non-verbal cues like eye contact and body movement are not lost, and text-based communication is worst because they are. Those non-verbal cues are some of the very things that we rely on for mentalization – our ability to build and maintain a working model of others minds – which empathy relies on.

Social presence can also be a function of people's expectations of a medium. It can also change with new design approaches. The use of emojis and avatars, for example, attempts to fill some of the gaps in text-based interactions.

There is a growing class of commercial services that try and engender a sense of social presence while also scaling the ability of professionals to interact with many more users. There are examples in healthcare, therapy, marathon training and education that use a mix of live and recorded video, messaging and voice calls, often based around an always-on account.

It is an emerging field, but surely the commercial pressure on many of these services will always be to automate more: maintaining the impression of a human relationship while hollowing it out with more automation and leaving only homeopathic levels of empathy. The problem is that, at least initially, users may find it hard to spot the difference.

Sherry Turkle has noted the ability of children to become attached to, and even to show empathy for, their Tamagotchis.* She sees one of the risks of the current generation of AI chatbots as tricking us into seeing empathy where none exists – not because of anything special about the technology, but because

---

* Tamagotchis were a short-lived 1997 craze involving tiny plastic eggs with LCD screens that needed regular 'feeding' and 'nursing'.

our brains are susceptible to this delusion. We *will* relationships into existence between ourselves and software because of how our brains are wired.[25]

This is a route that public services must not follow. The relationship that people say they need with public services does not lie that way. The aim of the designers of public services should not be 'the efficiencies of mere connection' but to anchor those services in genuine human relationships. The best medium for the combination of situation and relationship should be used, recognizing that sometimes it will not be a website or app.* Regardless of the medium, the role of digital should be to augment those relationships with better context. Technology should be in service of the relationship between the public and officials, not the other way around.

Universal Credit again provides a hint of a future model, and it brings to light some clear gaps. First, Universal Credit uses a combination of a digital account, in-person meetings and phone calls to communicate with users. So, situations that demand face to face can be done face to face. Second, the design of Universal Credit's digital account – and in particular the journal and to-do list mentioned in chapters 1 and 3 – were in part a response to the very public issues with Universal Jobmatch.

The Universal Credit journal put the relationship between claimants and caseworkers at the core of the service's design. Claimants and caseworkers use it to discuss job applications, resolve issues and ask questions of each other.† Because it provides a record of a claimant's interactions, it is transparent to both groups of users what has happened and when.

It also tried to create a shared space, halfway between 'citizen space' and 'government space'. The aim was to create not a one-sided government service telling a user what to do, but

---

* Also, that the 'best' medium will be different for different people.
† The Universal Credit journal is slightly different to the design pattern described in chapter 3 because it is both a record of what has happened and a two-way text-based chat interface.

instead a space for government officials and the public to work in – a space that was consistent between face-to-face, phone and text-based interactions, and one that could reflect the work done by either side.

But clearly, the Administrative Fairness Lab work shows there is something lacking. Some of that is likely the result of the politics of the welfare system and the use of administrative burdens as a policy tool. But some of it appears to be a result of the design values that the UK public sector prioritizes.

The lab also interviewed government officials working on the design and roll-out of Universal Credit in 2023. In comparison with the public, who prioritized the quality of the relationship, officials tended to prioritize clarity and ease of use. The public saw messages that were short and clear as lacking empathy, but officials saw their role as designing a system that was clear and easy to understand. This is an example of 'the service paradox' (see chapter 8), where the quality of a service as defined by professional standards results in suboptimal outcomes as defined by users of that service.

That Universal Credit prioritizes ease of use is unsurprising. The team of civil servants behind Universal Credit was probably the most mature digital team operating in UK central government in the 2020s. However, user-centred design, at least as it tends to be practised in the UK public sector, tends to be utilitarian.[26] The priority is getting the immediate task done and within the bounds of a particular service.

This bias towards utilitarianism represents a potential sinkhole lurking under the digitization of public services. When we use a digital service, it's not just helping us get things done. As Sherry Turkle puts it: 'The tools we use to think change the ways in which we think.'[27]

As the work of the Administrative Fairness Lab, Pamela Herd and Donald Moynihan shows, the public services that exist today are already forming people's understanding of the relationship they have with government. But they have been created in the absence of an understanding of what fairness looks like in public services, and with surprisingly little knowledge about what digital, as a medium, does to the relationship between the public and public institutions.

Universal Credit may hint at a few ways forward. Design for social presence and psychological ownership may also help. However, there is much that is new here. The design patterns in this chapter are best seen as places from which to start, rather than as being fuller answers. In time we may come to see that the answer is not digital at all – at least not directly. It might be that the best role digital can play is to free up more government staff to spend time sitting opposite the public they serve and looking them in the eye.

PATTERNS

1. Acknowledge the work of users
2. Show what has gone before
3. Anchor services in reality

Acknowledge the work of users

Members of the public *and* officials are users of services. Because some things are hard and time consuming for the public and some (potentially different) things are hard and time consuming for officials, the level of administrative burden experienced by the public and by the organization changes over time.

The diagram shows how the levels of activity change over time for members of the public and government officers as users of a means-tested welfare service. Because the outcomes of welfare services are a *co-production* between the public and officials, the level of activity of each type of user is linked.

Initially, the member of the public's activity (who we'll refer to as 'the claimant') is high. They are unemployed and, in return for welfare payments, are required to attend job interviews and training courses.

The activity of the 'case officer' is initially low. They meet with the claimant once and then review their progress digitally.

When the claimant's parents become ill, the requirements to find paid employment are reduced, so their activity gets lower. The case officer takes on the job of identifying part-time work that allows the claimant to travel to support their parents, so their activity increases. When a case officer finds the claimant a job, both of their activities reduce.

We could draw similar 'activity curves' for the efforts that farmers and inspectors put into a claim for agricultural subsidies, or for a patient with a long-term health condition and their health-care team.

The perception that burdens are fairly distributed between the public and officials can contribute to the quality of the relationship between them. Services should therefore surface the activity of the other side of a relationship (fraud investigations and safeguarding excepted).

There is no single way to do this, but it should always be clear to a user what activity the other side has completed since their

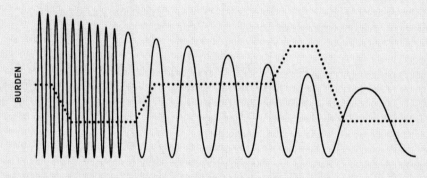

last interaction. Surfacing the effort that the other user is putting in, and acknowledging how this is changing over time, is a way to help users understand what is happening on the other side of the relationship.

Show what has gone before

Services should show respect for users by acknowledging previous interactions. While some services are one-offs, with a defined start and end, many are cyclical. The diagram compares one-off and cyclical services.

One-off services are hermetically sealed from everything that went before. A birth should only be registered once. A single vaccination may be enough to give people lifelong immunity.[28] A house is rarely sold by the same person twice.

In a one-off service, a user completes a set of tasks starting from scratch. Once the information is submitted, a decision is made about the user's eligibility and then there is some sort of outcome, such as money being sent or a credential being issued.

By comparison, cyclical services have a backstory. Passports are renewed, welfare claims and tax credits are reassessed. A referral to a specialist for a medical condition might be part of a series of ongoing assessments. As the diagram shows, tasks, decisions and outcomes are part of an ongoing loop.

To take an example, a service that assesses the needs of someone caring for a vulnerable adult would be a cyclical service. The service would need to understand what the carer might need in relation to financial support, help with housework, transport or access to respite care. This might include an interview, a home visit and a financial review.

The assessment would need to be regularly reassessed to see if additional support was required. Rather than starting from scratch each time, the service would start with what it knows and ask what has changed.

One-off service

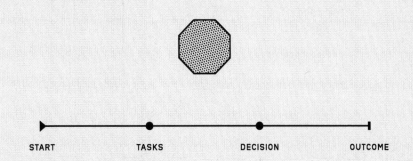

START   TASKS   DECISION   OUTCOME

Cyclical service

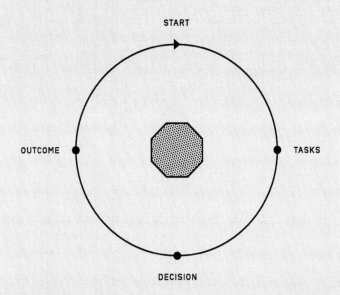

START

OUTCOME   TASKS

DECISION

Because that information is available to both the applicant and the social services officer conducting the interview, there is less chance that a user will feel misunderstood.

Anchor services in reality

Services should be anchored in the reality of the relationship between the public, services and government officials. This means digital interfaces should feel as though they exist in a space between the two sides of that relationship. Services should also be grounded in things that people recognize and that they can, hopefully, have some feeling of ownership over.

The diagram shows a user's account. Rather than being organized around services, it is organized around representations of things that are familiar to the user. These include their child's school, their car, their flat and their local hospital. Each thing has additional information and action associated with it. For example, a booking for a blood test appears next to the hospital.

The diagram also illustrates how user experiences can occupy a user's space, or the space of the service provider (labelled 'government space'). Where an experience feels like it lives is – to a large extent – a design choice. But if the aim of an experience is to mediate a relationship, then it follows that it needs to occupy a middle ground between the two spaces.

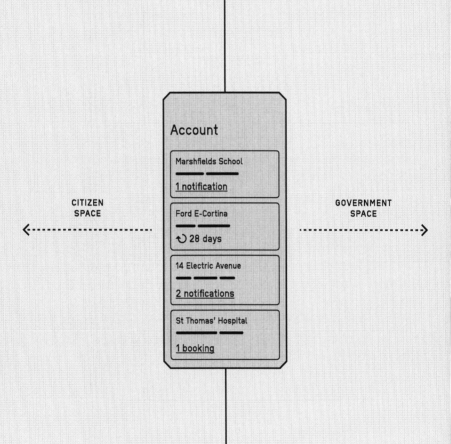

CITIZEN
SPACE

GOVERNMENT
SPACE

Account

Marshfields School
1 notification

Ford E-Cortina
↺ 28 days

14 Electric Avenue
2 notifications

St Thomas' Hospital
1 booking

STRATEGY

Foreground human relationships

Public services in the digital age are not digital-only services. They weave together the digital and the physical world.

If a service needs to 'add more human', technology should be an enabler. By operating online support, video, phone callback and the booking of face-to-face appointments as common components, it creates a more economical way for every service to add them as needed.

Real-world service discovery can be made easier, too. By maintaining data about what services are available at what location as a common resource, multiple services can draw on that data. For example, data about what health services are available, at what location and when the next available booking is can allow services to refer patients to the most convenient and timely provider.

Most governments operate some kind of public-facing offices: tax offices or job centres, say. These tend to offer only the services that are related to the part of government that owns them: you can't get help to fill out a passport application at a tax office. Government offices (local government, tax offices and labour exchanges) should also be treated as a shared capability for any government service that needs to talk to people face to face. Single government websites such as GOV.UK have been designed so that users don't need a mental model of how government works in order to use them; the same should be true of people's face-to-face experiences of government.

Task lists, journals and accounts (all operated as components) should be designed to show the hard work of public servants working on behalf of users.

It should also be easier to surface people and place within services. While there will be risks to displaying information about public servants, such as names and photos, there may be some contexts in which it is both valuable and acceptable.* Adding a photo of a government office or a hospital building that

---

* For example, it's possible to look up the names and photos of community police officers on the Greater Manchester Police website because it is considered important that people know who their local police officers are.

someone is due to visit should be less controversial. Again, these are things that can be solved once as a component.

Perhaps the biggest opportunity is in how services are designed. Public sector design practice needs to be given the space to consider the quality of the relationship that a service is creating. That will require leadership.

Main points

1. Empathy is understanding the feelings, thoughts and experiences of others. Digital services can block empathy.

2. Mentalization is the ability to build and maintain a working model of others' minds based on cues and context.

3. People don't just care about outcomes, they care about how things make them feel.

4. People place more value on the physical versions of products like photos.

5. Psychological ownership is the feeling people have that something is theirs.

6. Social presence is the sense of being with others and the sense that the relationship is truly represented.

7. Services that automate the representation of a relationship create fake empathy.

8. Acknowledging the work of users on both sides of a relationship may help create a sense that burdens are distributed fairly.

9. Services should explain the context of what has gone before.

10. Services can occupy 'government space' and 'user space'.

11. Services should be anchored in the reality of relationships and things that people feel ownership over.

12. Common components can help add human interactions to services.

13. Government offices should be operated as common spaces that can serve users of many services.

Examples: anatomyofpublicservices.com/examples/empathy

CHAPTER 8

Designing the seams (not seamless design)

On YouTube, there's a compilation of Steve Jobs speeches where he says: 'It just works. Seamlessly.'[1] There are forty-four examples in total. It's a phrase that's become a guiding principle for Apple.* If you watch an Apple keynote, you'll hear 'it just works' and 'seamlessly' (the 2023 Apple keynote, which launched their mixed-reality headset, had a 'seamless' count of eleven).[2]

'It just works' sums up Jobs's approach to design: remove and simplify. He thought design should 'get out of the way', as he said about the iPad and the iPhone, which were designed so nothing would distract from the screen.[3]

Jobs and his head of design Jonny Ive were inspired by the industrial designer Dieter Rams, famous for designing products for the German consumer goods manufacturer Braun. His maxim 'less, but better' and his '10 principles of good design' (which includes 'Good design is as little design as possible' and 'Good design is unobtrusive') have influenced generations of designers. The products created in Rams's time at Braun are beautiful, functional and pure. Be it a clock, a calculator or a music player,

---

* 'Seamless' is so associated with the company that, if the company launches a dud, people reach for it as a retort.

they represent tools designed around the essence of the task at hand, with little to distract.[4]

From the beginning Apple products were designed as tightly sealed boxes, to the extent that when Apple discovered third-party repair shops were opening the iPhone 4, they went to the effort of replacing the screws with tamper-proof ones. In part this was because Jobs didn't want users meddling with the purity of the product design, but he also believed that quality consumer products are tightly integrated and make good choices for users.[5]

In no small part because of Jobs and Apple, it has become an article of faith in the digital industry that a good design is one that just works. This is why Google's design principles include 'Simplicity is powerful' and Amazon's leadership principles have 'Invent and simplify'.[67] Simplicity is king. It's part of what makes those companies so successful.

One school of thought is that it is enough to apply Jobsian principles to the public sphere – to create public services that 'just work'. This contends that the types of services described in this book – ones that simplify complex interactions – are enough on their own terms, and that the instinct of the designer should be to abstract and simplify, if need be, at the expense of legibility.

But do we really want to design our public services like an iPad? Functional yes, even magical, but good luck if you want to understand how it works.

A functional, transactional view of the relationship between citizen and state – characterized by the argument: I pay my taxes, so all the council have to do is collect the bins and fix potholes – is hardly a new one. Some people are likely attracted to the idea that public services should 'get out of the way' on idealogical grounds, too.

But those arguments often come from the privileged position of people who don't *have* to interact with public services all that much. When they do, it's for relatively low-stakes things, such as applying for a driving licence. It also fundamentally mis-understands the nature of what makes public services *public*.

Public services are different because people *have to* use them. They *must* work, and they must work for everyone. They are also set apart from private sector services because their quality is harder to measure. If it's a service people *have to* use, you can't measure demand, price or profit – as you might with a phone.

Public services only get better through feedback loops between the public and the state. Regardless of how well they are perceived to have been designed by designers or policymak-ers, public services require an element of 'co-production' with the public to ensure they are of sufficient quality and can meet everyone's needs.

Aneurin Bevan, founder of the United Kingdom's National Health Service, is often quoted as saying the following about the service's creation:

> The sound of a dropped bedpan on the floor of Tredegar hospital should reverberate in the Palace of Westminster.

This has become the go-to shorthand for ministers taking responsibility when things go wrong with public services. But it appears Bevan was actually talking about something different: the right to complain and be heard that is created by a *public* service. He believed that people had complained previously, but they had not been heard. He said that with the creation of the NHS, 'a public megaphone will be put in the mouth of every complainant so that it can be heard all over the country'.[8]

The ability for users to 'meddle' is a key feature of public ser-vices. Democracy is about more than voting every four or five years, it's about the opportunities people have to shape the ser-vices and the rules that, in turn, shape their world.

One feedback loop for digital public service is user research: interviewing service users and testing alternative designs to understand users' needs. In 2013 the Government Digital Service introduced the Digital by Default Service Standard, a baseline that all digital services were expected to meet. We included a requirement that digital public services do ongoing testing with the users of services, and that services should be resourced so they could be improved on a 'very frequent basis, in response to user feedback'.[9] The result was that most UK government digital services are tested with real users throughout their design and development and are better for it.

However, despite the aspiration for user research to be an ongoing activity, it has tended to be an upfront activity, used for developing new services and features. Universal Credit, GOV.UK and a handful of other services succeeded in making it 'business as usual', but even done well it's a feedback loop on government's terms: service delivery teams choose the areas of focus, or, more often, the realities of business cases determine the focus.

That's not a criticism of user research: the situation cannot be any other way. Resources are limited, as is the ability of designers and researchers to ask the right questions. It's just not possible to understand every need from the centre. User research is an absolute must in the development of digital public services, but it is insufficient on its own as a form of co-production.

The concept of co-production has its origins in the early 1970s and the work of Elinor Ostrom.[10] Ostrom's lifetime of work on the subject was informed by fieldwork with communities who relied on shared, common resources such as fisheries, watersheds and public services. In one example, Ostrom found that larger consolidated police forces reduced the quality of policing (as defined by the public) compared with smaller, locally accountable departments, which had greater opportunities for public involvement.[11]

Ostrom also described the 'service paradox'. As noted in chapter 7, the service paradox is where the quality of services as defined by professional standards results in suboptimal outcomes as defined by users of that service. To give a specific example, better-designed textbooks might actually make education worse if the content is so clear and well designed that students no longer feel the need to discuss issues with their class or teachers.[12] This is because, in a school, pupils are co-producers of learning with their teachers and with each other. It has also been shown that students actively seek copies of textbooks whose margins have been annotated by previous students.[13]

Ostrom's theories explain that the effective management of a common resource, such as a public service, is a function of the interplay between governance at different *levels* (national, regional, community) and at different levels of *formality* (laws, regulations, social norms). Urban planning, for example, is not only a set of centrally set rules: local norms, the opinions of conservation groups, the style of previously permitted developments, and how a community uses a space all play a role too.

The management of public services is dependent on there being opportunities for different groups to take part. Indeed, democracy has been called the 'co-production of rules'.[14]

Co-production is a function of the *use* of services, not an upfront or periodic design activity from the centre. Services have to be designed to enable it. Ones that just work, seamlessly, don't make much space for co-production.

To avoid the service paradox in the next generation of public services, there need to be clear opportunities to understand the workings of those services. That's because understanding the way things are is a precondition for being able to change them. Democracy is, as the former British prime minister A. J. Balfour said,* 'government by explanation'.[15]

At the Government Digital Service, our version of 'it just works, seamlessly' was 'do the hard work to make it simple'. It was one of ten design principles published in 2012. More than

---

* Or it might have been a much later cabinet minister, Geoffrey Howe, that said this – no one is too sure.

any of the others, that principle summed up what we'd tried to do with GOV.UK: people should not need to understand government to interact with it; for all the reasons discussed in this book, using technology and design, we could shelter users from the complexity of government services.[16] However, as the service paradox shows, it is possible to make things *too* simple.

In designing GOV.UK this is something we didn't acknowledge loudly enough.[17] Despite talking about making things simpler, clearer and faster, at the same time we made some things harder to understand. GOV.UK simplified people's interactions with government because there was one place for people to go, rather than 2,000 different websites.[18] But it also obfuscated something of how government worked. It made it harder for people to understand which rules and organizations they are interacting with.

In a 2019 Gresham College lecture about the digitization of a benefit appeals process under the banner of GOV.UK, the legal commentator and journalist Joshua Rozenberg pointed out a consequence of this:

> One flaw in the current system is that the courts and tribunals website is currently headed GOV.UK on every page. In previous Gresham lectures I argued that this was inappropriate for claims against the government. Courts must be seen as independent of the parties. That argument seems to have been heeded at last... We wait to see whether the URL – the 'address' of the web page – will change too.[19]

There was no way, it seemed, for users to orientate themselves.

In physical space, people orientate themselves by 'way-finding', based on cues from the environment. Those cues come from things such as the paths we travel along, the boundaries between districts that we recognize, and prominent landmarks.

The term way-finding comes from the 1960 book *The Image of the City* by Kevin Lynch.[20] He defines 'imageability' as the extent to which a place invokes a clear mental image in people's minds. The imageability of a city determines how people orientate themselves and how they navigate.*

Services such as GOV.UK and the ones described in chapter 2 abstract away the structure of government. They are the equivalents of low-imageability cities. They provide few cues about where people are and who they are interacting with.

The example of the court service should never have happened, and it's not clear how it did. Early on, the GOV.UK team made an effort to ensure that government agencies with independent oversight functions maintained their own digital presence. There was also, initially at least, clearer separation between government as an agent of service delivery and government as a political institution.[21]

Working on Universal Credit in 2014, it was abundantly clear that the next-generation digital public services that automate, abstract and have complex data flows demanded a different approach to design. The development of the Universal Credit journal and task list were in part an acknowledgement that digital services needed to explain themselves better (see chapter 1). Later, when the Government Digital Service was beginning to think about the implications of common components, we again explored how to surface more of the structure of government.

But as design practice spread across government, the simplicity principle took on a life of its own. It developed into what, at times, felt like a tyranny of design, where anything that distracted from the proximate user need was impossible to justify. The idea that people should not have to understand the rules and the structure of government to get something done seemed to have morphed into an assertion that the structure of government and the rules should be obfuscated.

---

* In the book Lynch compares the high imageability of Boston, with its winding backstreets, the monoculture of New Jersey and the sprawl of Los Angeles.

Over a decade after the Government Digital Service design principles were published, it's clear that the public sector design practice that emerged in the United Kingdom from the 2010s onwards was fundamentally incomplete at inception. A Jobsian view had won out: that public services should work seamlessly, the instinct of the designer should be to simplify, and users should have minimal opportunities to understand the workings of services.

How then should we reconcile the fact that digital collapses the distinctions between organizations, and that the reuse of data is inherently opaque, with the need for people to understand how society works? Is there a fundamental contradiction here between digital-age design practice and democratic principles, or can we reconcile 'it just works' with 'government by explanation'?

While Dieter Rams's designs may provide the quintessential examples of functional simplicity, his view of functionalism is broad. It includes things like repairability, environmental impact and what we can call the 'understandability' of products.

In a speech in 1980 he asked designers to ask of their work:

Does it arouse curiosity and the imagination? Does it encourage desire to use it, understand it and even to change it?

And alongside 'as little design as possible', another of Ram's '10 principles of good design' is that:

Good design makes things understandable – it clarifies the product's structure. Better still, it can make the product talk. At best, it is self-explanatory.

What might it mean to make digital public services explain themselves – to make them encourage understanding and change? The answer is not that we should abandon simplicity, but that it must budge up and make a little space. Rather than aiming for seamless design, the aim should be for simple services that wear their seams proudly.

For inspiration, we can look to the work of Mark Weiser, who was the chief technologist of Xerox PARC in the late 1980s and the 1990s. In a 1994 essay titled 'The world is not a desktop' he railed against the emerging metaphors for computing: terms such as 'intelligent agents' and 'virtual reality' – not because they distracted, but because they were an *imposition*. They made themselves 'too much the centre of attention'.[22] A good tool was, he said, 'an invisible tool'. He explained:

> By invisible, I mean that the tool does not intrude on your consciousness; you focus on the task, not the tool.

He argued that we should aspire to design 'calm technology'. 'Beautiful seams' would, he proposed, be the way to interact with digital tools that would otherwise exist in the background.[23] Rather than being totally hidden, complexity is there to be revealed as needed. Users can configure, understand or take control of automated processes, as needed.[24]

It is an approach, as researchers Matthew Chalmers and Areti Galani note, that is suited to everyday interactions in which a user's attention is on getting a task done but whose underlying infrastructure is effectively invisible, despite its effects being very real ('literally visible, effectively invisible'). As such, it suits public services perfectly.[25]

Weiser died before he could fully explore the idea. There are examples of his vision being brought to life, though. Digital thermostats that can reveal the schedule they created are one example. The visibility of URLs in web browsers and the ability to 'view source' to reveal the underlying HTML code have been proposed as others by web developer Jeremy Keith.[26]

Public services are not web browsers or thermostats, but seams exist between all the anatomical parts described in this book. They exist service to service and organization to organization, as the GOV.UK courts service example shows. Showing something of those seams is a way of co-producing a public image of how democracy works and where power and accountability lie.

Seams exist between databases, and harms can occur when data moves across them. In 2018, for example, the UK National Health Service pulled out of an agreement that granted Home Office immigration officers access to data about patients. The withdrawal followed objections from public health officials and parliamentarians that people were not seeking medical attention or were deliberately missing vaccinations for fear of data about them being shared.[27] People will have valid reasons for fearing how information is used, and not all of those reasons will be apparent to the people designing services.*

In Estonia, the Andmejälgija ('data tracker') service provides a way for users to understand how data about them is used. They can see which organization has accessed what information, when that information was accessed and for what reason. Only a fraction of the public use Andmejälgija, but it creates an expectation from the public that data will not be misused and a clear understanding by public servants that misuse will be discovered.[28]

Putting users in control of how data about them is used means they become the best protection that government has against fraud and that citizens have against misuse. Data protection and fraud prevention can be co-produced with the public.

There are also seams between services and rules. Digital services put decisions at a distance from users and make the rules opaque. In 2022 the Bingham Centre for the Rule of Law published a report into the use of technology for things such as contact tracing during the Covid-19 pandemic. The report made the point that there is a design challenge to explaining the law while also accommodating a proximate outcome for users:

---

* Most of us are fortunate enough to be able to safely ignore the ingredients list on a food packet, but if you have an allergy it can be a matter of life and death so governments mandate that they highlight any allergens in bold text.

It is about helping people understand the situation that they are in, what rules apply to them, what the consequences of these rules will be, and what kind of behaviour to take in response to these rules. This type of communication design is a legal design challenge.[29]

Anything can become a conspiracy when you don't know how it works.[30] Providing routes into understanding the rules that underlie a service enables the co-production of those rules and public understanding.

Finally, seams exist between the public sector and the rest of society. Advocacy groups need ways to shape the rules too, especially where a trusted relationship does not exist between the public and government. The Child Poverty Action Group's 'early warning system' is one such example. It collects first- and second-person experiences of using the United Kingdom's welfare systems and then uses them to advocate for specific design changes to policy and digital systems.[31] Linking to advocacy and community groups at the point of use enables those groups to play their part in co-producing better rules and better services.

With hindsight, the Government Digital Service's '10 design principles' from 2012 need a couple of additions:

11. Show the seams.

12. Democracy is a user need.

Public service design requires a reset. Creating services with beautiful seams (or we could say *democratic seams*) provides a way of revealing decisions, rules, data and accountability to users while not abandoning the aspiration for simplicity. Superficially, it may appear that they are less utilitarian or functional or minimalist, but so be it. A public service is not an iPad.

PATTERNS

1. Help users orientate themselves
2. Explain when data will be used in a new context
3. Explain decisions progressively

Help users orientate themselves

Services should orientate their users, allowing them to adopt the correct stance for the organizations they are dealing with. How a user feels about interacting with a healthcare provider, for example, may be very different to how they feel about interacting with the immigration system. Users should also be able to find out why a service works the way it does and how they can change it.

Services should help users understand, at the point of use:

1. which organizations they are interacting with;

2. where accountability lies;

3. where to find the underlying policies and legislation;

4. what opportunities exist to shape the service; and

5. which community organizations can help them.

The diagram shows two services. The first is for managing claims for agricultural subsidies based on how the claimant's land is used. It includes links to the government minister responsible for the service, the blog of the operational team, and the underlying legislation and policies. The second service, which is operationally independent, allows farmers to appeal decisions made by the first service. When moving from one service to the other it is clear that they are now dealing with a different organization.

In the first service there is a link to join a service user group that gathers feedback on the service and regularly tests new features and processes. In addition to links to the official consultations and user groups, there is also a link to a local farmers' group and a national group who operate an 'early warning

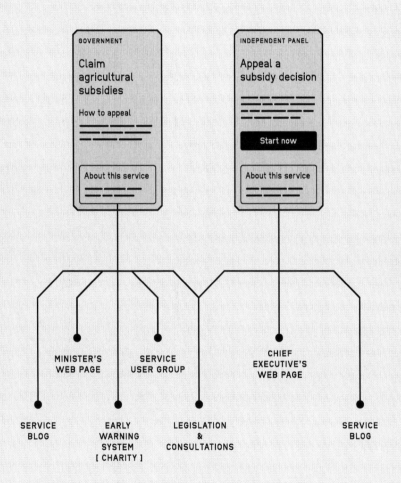

system' for the digital service. Both services link to the underlying legislation for agricultural subsidies. When changes to the legislation are proposed (e.g. how regularly payments are made), users can click through to take part in a consultation.

None of this information should be so prominent that it gets in the way of a task a user is trying to do, but it does require compromising on minimalist design. However, without making accountability and agency clear, services will remain products of the organization that created them rather than being a co-production with their users.

Explain when data will be used in a new context

Digital services should help people understand how data about them is being used, at the point of use of a service. When data is used in a new context, people should be told.

The diagram shows a user registering for a local property tax after moving into a new home. The local property tax service gives them the option of telling the driving agency, the pensions agency and the benefits agency that they will be living at the property. In this example, notifying the agencies is optional. They can choose to do it separately (e.g. they might not be moving into their home immediately).

The user's pension and driving licence are updated immediately. However, because moving house might have financial implications for how their benefit payments are calculated, they receive an additional notification. This explains how their payments will change and gets them to confirm the date from which they will be living at the new address.

Explaining when data will be used in new ways creates friction, but it is necessary friction. That's because the once-only principle means that the cost of data being wrong is much higher and the underlying data is a bigger target for fraudsters. Notifications like this are an opportunity for users to confirm that it was them who reported a change rather than

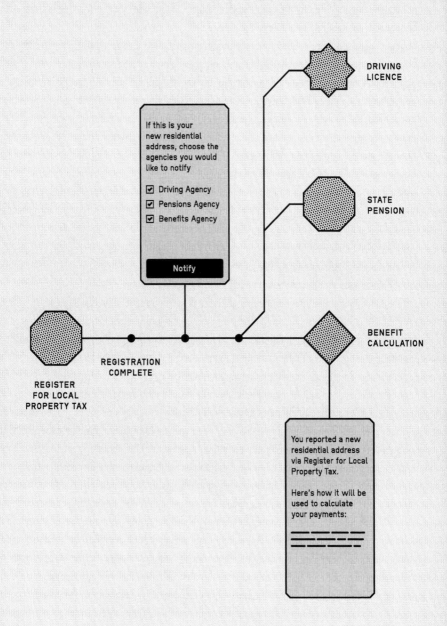

DRIVING
LICENCE

If this is your
new residential
address, choose the
agencies you would
like to notify

☑ Driving Agency
☑ Pensions Agency
☑ Benefits Agency

Notify

STATE
PENSION

REGISTER
FOR LOCAL
PROPERTY TAX

REGISTRATION
COMPLETE

BENEFIT
CALCULATION

You reported a new
residential address
via Register for Local
Property Tax.

Here's how it will be
used to calculate
your payments:

someone else doing so fraudulently. Furthermore, the risks of data bubbling up in unexpected places are not evenly distributed across society. Someone fleeing an abusive relationship, for example, has good reason to be concerned about their personal information appearing in a digital space that their partner has access to.

Giving users a knowing understanding of how data about them is being used creates millions of engaged curators. They are the best protection government and the public have against fraud and misuse. In addition, as common data infrastructure makes it easier to join data together, what data gets joined together for what purposes becomes a question of political judgement rather than one of technical constraints. Helping users understand those joins supports a better public debate about those judgements.

In these ways we can see that anti-fraud activity, data governance and trust are all things that are co-produced with the public.

Explain decisions progressively

Users of public services should be able to reveal how decisions have been made, at the point of use. This applies to decisions made by public servants and by automation.

When someone is using a service and something goes wrong, or something just doesn't look right, they have to switch task from seeking an outcome to trying to understand what has happened. Filling in a form or checking a message is a fundamentally different task to understanding how the underlying system works. They exist on a different plane, so there need to be gateways between them.

Rather than overwhelming users with information, services should use progressive disclosure to explain decisions bit by bit.[32] At each step users should then have the opportunity to return to their original task or find out a bit more.

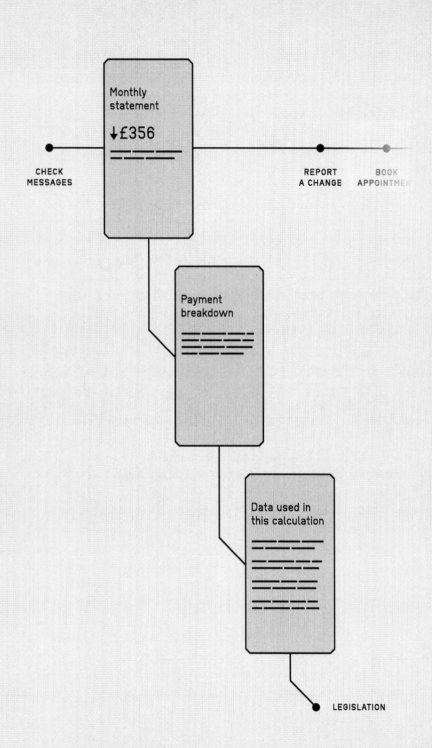

The diagram shows a user signing into their account to report a change in their rental costs and to book an appointment with an advisor. They notice a new message about their upcoming payment and see that it is lower than expected. They abandon their initial task and click through to see a breakdown of the payment. It shows how the payment total has been arrived at. They notice that the amount of money they received for childcare and housing costs is lower, but they still don't know why.

Because the statement still doesn't make sense, they click to see the exact data that was used by the calculation rules. This means they can see what data was missing and was therefore *not* used by the rules. They take this information to an advisor who crosschecks the missing data with the policy rules and helps appeal to the service for the statement to be recalculated.

Giving users opportunities to understand the rules in this way means circumstances not foreseen by policymakers and designers can be identified much sooner. It creates opportunities for users and the people who support them to co-produce *better rules*.

Digitization can marginalize the decision-making process, putting it at distance or hiding it from direct view.[33] But it does not have to be that way. Digital services can put explanations within reach of users, in ways that are impossible for analogue services.

STRATEGY

Put transparency at the point of use

Design practice in the public sector needs to expand to include understandability and explanation, while not abandoning simplicity. Digital public services shouldn't *just* work, they *must* work. They must also explain themselves, because public services only get better through feedback loops and co-production.

Putting transparency at the point of use and applying the principles of seamful design means it is possible to design services that are simple to use but that also actively educate people about how their democracy works and where power and accountability lie. This, in turn, can give people ways to understand and shape services. It also creates a 'salience of responsibility',* as Rob Kling (see chapter 1) put it.[34]

Many of the changes required to put transparency at the point of use may actually be quite small. Just enough information needs to be added to acknowledge the reality of the rules, decisions, data flows and accountability mechanisms that already exist but are hidden.

Nonetheless, seamful design in public services represents a cultural, legal and design challenge. There is a cheap comfort that designers can take in the purity of designing for the task at hand – just as there is cheap comfort in legal professionals or policymakers being able to point to the totality of a law or a policy document as the definitive truth. The hard work lies in creating an effective working practice between the professions.

A new design philosophy for the public sector needs to be forged, and with it a new design coalition that includes legal and policy professionals. Digitizing public services in the absence of those things points us towards a future that is, frankly, not very democratic. But get this stuff right and public services can work much harder for users while also allowing users to be effective co-producers of those public services. By putting transparency at the point of use – by public services wearing their seams proudly – public services can get better for everyone.

---

* In the 1970s Kling proposed that software engineers' names should be printed directly on manuals as a way of creating some accountability.

Main points

1. Seamless design and simplicity is a false target.

2. Public services only get better because of feedback loops; they require an element of co-production with users.

3. The 'service paradox' is where the quality of services as defined by professional standards results in suboptimal outcomes.

4. Democracy is the 'co-production of rules' and 'government by explanation'.

5. Digital services and infrastructure can make it harder for people to understand which organizations they are interacting with.

6. Users should not have to understand the structure of government, but that does not mean it should be obfuscated.

7. People will have valid reasons for fearing how information is being used and not all of these will be apparent to people designing services.

8. Millions of engaged curators are the best protection governments have against fraud and that citizens have against misuse.

9. With seamless design, complexity is there to be revealed as needed.

10. Explain when information is being used in a new context.

11. Trust users to control how data about them is used at the point of use.

12. Users should be able to access historical views of how data has been used.

13. Help users understand decisions.

14. Public sector design needs resetting.

Examples: anatomyofpublicservices.com/examples/seams

CHAPTER 9

Accountable automation

In 2023 NASA's Jet Propulsion Laboratory sent a software update to the two *Voyager* spacecraft. Forty-six years into their missions, they were the only human-made objects in interstellar space. The spacecraft's thrusters, which keep the communications antennas pointed at Earth, had begun to clog up with propellant. The software update provided a fix: it would reduce the number of times the thrusters fired during each manoeuvre, prolonging the life of the mission.

The craft were designed to be updated in this way. In-flight reprogramming had begun with the earlier *Mariner* missions to the inner planets, and this process was routine by the time of the *Voyager* launches. As the craft passed the gas giants of Jupiter and Saturn, a software update improved how the images the two craft took of those planets were compressed, meaning they could be sent back to Earth more quickly. *Voyager I* was updated a total of eighteen times during its Jupiter encounter alone, and both craft continue to be updated throughout their missions.[1]

These updates were possible because of an affordance of software: mutability. Mutability is the degree to which something can be changed. Compared with something physical, such as a bridge, digital products and services are *highly mutable*.

Purely physical things don't change much after they have been designed and built – not radically at least (a suspension bridge does not suddenly become a cable car). With digital things, what they do and how they do it can radically change over time. They have the potential for change *after* they have been designed.

Instagram started as a mobile check-in app before pivoting to photo sharing and then gradually gaining all the features of a social-networking app as it tried to capture a larger userbase.[2] Similarly, Apple enabled humidity and temperature sensors in their HomePod Mini smart speakers after launch.[3] While in the early days of app stores, it seemed common for each new update of an app to request access to more and more sensor data: a torch app suddenly wanted access to an exact GPS location; a game wanted full access to contacts.

In a way not readily apparent to people working in close proximity to the creation of software, most software is in a state of *constant* flux. Many of the concerns of digital rights campaigners can be understood through mutability. The fundamental worry is that, because of the 'latent capabilities' of software, the products and services we use today could be quite unfamiliar tomorrow.

The misconception that software is mostly static, most of the time has been perpetuated by software vendors trying to sell 'solutions', especially in the public sector. The idea that buying a software system is the same as buying a new fleet of buses or commissioning a new school building became a comfort blanket for public officials searching for certainty. If there is an understanding of change, it is often wrapped up in constructs like 'versions' or 'phases'.

People may like to think of digital systems as static, but they are stable in the same way that a river is stable. The River Thames feels like a constant to a Londoner.[4] It's always there, and its name reaches into prehistory.* But speak to a hydrologist and they'll describe a chaotic system that is in constant flux; speak to a palaeontologist and they'll point to the bones of the

---

* 'Thames' is probably the second oldest named thing in England.

ice age megafauna – woolly rhinos, mammoths and lions – that once walked its banks, now lodged in its gravelled terraces.[5]

Digital systems are – like rivers, like species – in a near constant state of change, and computer science has its hydrologists and palaeontologists.

Meir 'Manny' Lehman was working on a side project at IBM in the 1960s when he discovered that, despite everyone's assumption that they were 'done', the maintenance of computer operating systems was sucking up more and more of people's time.

Lehman's initial work was based on a study of changes to IBM's OS/360: the operating system for IBM's mainframe computers that were in common use (or as common as any computer was back then) in the 1960s and 1970s.

When he left IBM Lehman continued his work while working as a professor at Imperial College London. Throughout the 1970s, 1980s and 1990s he and various colleagues developed a set of theories – known as the laws of software evolution – about how and why software changes. The first law is titled the 'law of continuing change'.[6] Not only does software change, Lehman and his colleagues argued, it needs to change. More than that, it needs to *evolve* in response to its environment.[7] Most software does not solve a tightly defined problem or implement a specific algorithm that can be specified up front. It exists in a real-world environment, so it is strongly linked to changes in that environment.

Just as geology exerts evolutionary pressure on rivers, and the ebb and flow of ice sheets exerted evolutionary pressure on the woolly rhinos of the Thames, so users exert evolutionary pressure on software.

The evolution of software means that upfront software design is essentially *predictive* (or, less charitably, a guess). The 'correctness' of software is determined by user satisfaction and the fit with its environment rather than by 'equivalence to a specification'. It is this evolutionary pressure that results in 'a process of never ending maintenance activity'.[8]

Today, the whole practice of software development has become organized around the change that Lehman and his colleagues identified. Software engineering, design, user research and even business development are arranged around short iterative cycles and the management of ever-shifting complexity.

For some projects, the scale and pace of change is very high. Between the end of February 2024 and the end of March 2024, the source code of the core part of the Linux open-source project had 616,984 additions and 184,323 deletions, with changes made by 1,042 different people.[9] Linux is a huge project, but even modest digital teams will update their services weekly or monthly, and some will deploy changes hour by hour.[10] Staying on top of this rate of change is only possible because of specialist tools for version control, automated testing, the management of dependencies on external software libraries and managing deployments.

Agile project management organizes work around small teams who work in weekly sprints to deliver working software. Systems are engineered with changeability in mind, applying the idea of 'small parts loosely joined' (see chapter 5), so parts of a system can be swapped in and out. User researchers and designers test alternative designs with users and changes are made to the live service based on what they learn.

As with the example of Instagram, businesses may even pivot their whole focus based on the data they have about how users are behaving.[11]

It's all an ongoing flow of change. But as more public services become digital public services, this 'changeyness' creates an accountability gap. The people we rely on to hold public services to account increasingly lack the ability to do so, because of a lack of 'ground truth'.

If an elected representative walks into their local hospital or a journalist visits a local government office, we can be fairly confident that they will experience the same physical space as when they visited the week before. They can make a reasonable guess about what questions to ask and of whom. But this is not the case with digital services, which, at times, seam to defy integration.

As immigration solicitor Jonathan Kingham noted in an article on the digitization of Brexit-era immigration systems in the United Kingdom:

> Unlike with rules and legislation changes, there was little opportunity to scrutinise the detail of what are, in fact, significant changes to the immigration system prior to their coming into force (bar selective 'user testing', which is rarely transparent to all).[12]

In India a system called DigiYatra is increasingly used by airlines and airports to implement paperless travel using facial recognition to identify passengers. Initiated by the Ministry of Civil Aviation, as of mid June 2023 nearly 1.75 million people had travelled 'seamlessly' using the system. But civil rights lawyers have raised concerns about the lack of oversight and transparency over how data is handled.[13]

When the charity Child Poverty Action Group was researching Universal Credit, the United Kingdom's digital welfare system, they found there were few public sources that explained how the service worked.

The UK government cited constant changes to the digital system, along with personalization, as reasons for not releasing screenshots. As a consequence, researchers had to do the equivalent of sending a courtroom artist to sketch proceedings, recreating how parts of the service worked by creating mockups of screens and messages. Through this approach they eventually identified questions that were missing from the application process, meaning that some people may have been unable to claim their full welfare entitlements.[14] Similarly, on RightsNet, an online discussion forum for welfare rights advisors, one user recreated their own version of the Universal Credit application process using Google Forms, based on screenshots they had collected.[15]

As the UK government noted in its response to the Child Poverty Action Group, digital services don't just change over time, they also change *between users*. They use the data they hold about people to change each user's experience. While two people visiting the social security office will occupy the same physical space, this is again not the case with a digital service. A service might, for example, use the fact that it knows someone's income to suggest welfare benefits they will likely be eligible for, or perhaps it will use their location to send them notifications of planning decisions.

The use of data in combination with software in this way is inherently opaque and further adds to the task of understanding how digital public services operate. Add to that the fact that some software (particularly AI-based software) is non-deterministic, so the same user may get different results on different days, and the task of understanding is larger still.

This is not an entirely new problem, either. Software bugs dating from the 1990s were responsible for millions of people (mostly widows, divorcees and women who rely on their husband's pension contributions) being underpaid their UK state pensions for decades. The problem affected 134,000 people and the average underpayment was estimated to be £8,900. The issue had been known about since the 1990s but was not made public until 2022. Government ministers claim that even they were not told about it.[16]

The reality that campaigners, parliamentarians and lawyers find themselves confronted with is often akin to that faced by Arthur Dent in the opening chapter of *The Hitchhiker's Guide to the Galaxy*.[17] Standing in front of a bulldozer trying to find out from a local government official why his house is about to be demolished, Arthur is told:

'But the plans were on display ...'
'On display? I eventually had to go down to the cellar to find them.'

'That's the display department.'

'With a flashlight.'

'Ah, well the lights had probably gone.'

'So had the stairs.'

'But look, you found the notice didn't you?'

'Yes,' said Arthur, 'yes I did. It was on display in the bottom of a locked filing cabinet stuck in a disused lavatory with a sign on the door saying "Beware of the leopard".'

Part of the comedy in Arthur Dent's predicament is that we know he *should* have been told. In a rule-based society, a lot of government activity is dedicated to putting dull but important information on the public record in a systematic way.

In his book *The Rule of Law*, Tom Bingham, a former Lord Chief Justice of England and Wales, listed eight principles that define the concept of the rule of law. The very first principle reads: 'The law must be accessible and so far as possible intelligible, clear and predictable.'[18] For this reason, tax rules and policies are published for public scrutiny. Notices are posted in official gazettes. Public registers are maintained for land ownership and company ownership. Planning notices are published (without the deterrent effects of leopards). We see this repeated across different aspects of public life. It helps make society legible.

The reality of how a service is operating and how it is changing is only truly encapsulated in the service itself. As Jonathan Kingham – quoted above about the changes to the post-Brexit immigration system – went on to note, digital services 'increasingly take centre stage' over the law that underpins them. As such, the implementation choices of designers and software engineers have become critical in how the public access things that they have a legal right to access.

Why then is information about how services and digital infrastructure work so difficult to come by?

While the UK government may have cited the constant change of the Universal Credit system as justification for not publishing the design of its screens,[19] that is a bit like saying we can't create a map because rivers move. Maps are never a 100% accurate representation of reality. They are, by definition, an abstraction, a snapshot in time. Government mapping agencies have to conduct ongoing surveys to keep the maps relevant and at their most useful (although even a slightly out of date map can be helpful).

Maps are an example of what is termed a 'boundary object'. A boundary object is something that serves different purposes for different groups but has enough common truth to be useful to all. Boundary objects help different groups to understand each other.

A building developer, an environmental scientist and a lawyer might all use the map of an area, but they would do so in different ways. The developer would use it to show where a proposed development would be built, the environmental scientist to understand the impact of the development on the local ecology, and the lawyer to advise a client at a nearby property about the impact on their rights of way.

Biological 'type specimens' are another example of boundary objects. According to the Linnean Society of London's definition, a type specimen is 'a specimen which is permanently associated with a given scientific name, and acts as a permanent reference to confirm the identity of the species'.[20] There is a duck-billed platypus in the Natural History Museum in London that, for the purposes of scientific description, defines *all* duck-billed platypuses.[21] The fruit fly is probably the most studied animal on the planet, but it is defined by a specimen caught in 1830.[22] Type specimens can't represent the full genetic diversity of their species, but they provide a stable reference point for everyone to understand that they are talking about the same thing. Boundary objects exist in the public sector too. Land registration documents, company registration records and planning notices all act as such.

The boundary objects for digital services and infrastructure are the things that provide ground truth about how they work and how they are changing. Luckily, the tools and practices that have grown up around software development mean that much of this information already exists. Some of these have already been mentioned, including the version history of code and automated testing. Reviewing changes to the software code for a service is one way to enable public scrutiny. For this reason, when we created the United Kingdom's Digital by Default Service Standard, it included a requirement for services to publish their code.*

Automated testing provides another opportunity to explain how a service works. Software tests are code that checks code, written by software engineers to make sure things work as they should. There is a saying in software engineering: 'If you didn't test it, it doesn't work.'[23] A typical example of a software test might look like this:

```
GIVEN a user is logged in
WHEN they click on the 'my account' link
THEN they can view the list of services they are using
```

This describes how a feature of a service *should* work. These sorts of test can be written to be readable by a non-technical person but also to be run automatically by a machine.

In addition to code and tests, there are other byproducts of software development that could act as boundary objects.

When a service is updated, the team may create release notes that describe significant changes. If there is a security issue, they will create an incident report. There may also be what is called a 'reference model'. These are also a description, in code, of how something is expected to work. The European Union, for example, published a reference model for the European Digital Identity Wallet that can be used both by different member states to implement their version of the wallet and by civil society groups who want to understand how it will work.[24]

---

* Sadly, as time went on, it was rarely enforced.

There are processes for how data is managed too. Teams need to keep a record of the different databases they use and the data they store in them. In Estonia the government has taken that information and made it public. RIHA (Administration System for the State Information System) describes the what and the why of public sector data in Estonia. The entry for the Estonian population register, for example, details the different data types that are maintained for each citizen. It links to the legislation that the register operates under, explains that the Ministry of the Interior is the custodian, and provides a time-stamp for when the data was last updated.[25] Estonia also pub-lishes cryptographic proofs that its databases have not been tampered with, which can be used by auditors.

In addition to code and data, how services are designed creates potential boundary objects too. Part of user-centred design practice involves maintaining a list of user needs. When we launched GOV.UK as an alpha, we had an internal tool called the Needotron for keeping track of the user needs we had identified. It was the definitive source that designers and policy officials could refer to. Each need linked to individual pages on GOV.UK so we could see all the ways the need was being met. Eventually, that information was published in the open. In an example of seamful design (see chapter 8), if you added /info to the end of any gov.uk url, you could see the user needs for that page too.*

User-centred design teams create many different prototypes of potential digital interfaces. The United Kingdom's Depart-ment for Education publishes its prototypes online as a 'design history' that explains their design decisions.[26] Teams may also generate screenshots of how interfaces look on different devices and recordings of how they sound in screen readers.

Publishing information about how public services work and how they are changing should not be seen as transparency for

---

* This was later removed for reasons unknown.

transparency's sake. The creation of digital services is organized around change. Accountability mechanisms need to be organized around change too.

Effective public services are not solely the creation of policy officials or designers. They must be a co-production with campaigners, lawyers and parliamentarians (see chapter 8). There must be a competition for ideas over what is working and what is not.

E. M. Forster's 1909 short story *The Machine Stops* imagines a world in which humanity has become so reliant on an intelligent machine that it is incapable of public debate about its rights and wrongs. Then, when the machine starts to break, people are incapable of fixing it. We are currently a long way from that scenario, but it is fair to ask: as the public sphere becomes more digitized, what does it mean for society if we have no ground truth about how public services work and how they are changing?

The highly personalized worlds built for us by social media platforms have led to a lack of shared public experience of current affairs. The changeyness of digital services risks the dilution of the shared public understanding of how public services function.

I remember when the photo-sharing website Flickr launched its 'interestingness' algorithm in the mid 2000s.* Every photo got a score, which Flickr used to show more interesting photos to its users.[27]

To the eyes of the 2020s, Flickr's 2005 community forums read like a checklist of today's concerns about algorithmic decision making. How does it work? Is it fair? Can it be gamed? What does it mean for localized or cultural variation? These questions were asked more with curiosity than alarm, but today we ask them about automated decisions in welfare systems, policing and elections rather than about how holiday snaps are sorted.

---

* I was working at moo.com at the time, and we built a digital printing service on top of Flickr.

Back then it just seemed like a niche internet thing – in the days before the internet became everything.

In his book *Human Compatible*, on how to mitigate the risks of AI, computer scientist Stuart Russell suggests that everyone should also have the right to mental security to live in a largely true information environment.[28] Digital systems, whether infused with AI or not, create gaps in our public information environment. They create information asymmetries, where the teams operating services and infrastructure have a fine-grained, real-time view of how services work and how the public are using them, while those who need to hold the government to account for its actions operate largely in the dark.

As all our public services become – to a greater or lesser extent – *digital* public services, there will need to be a largely true information environment about how services and infrastructure work.

The ultimate aim of transparency should be better public services and a broader debate about what constitutes 'better'. For that, we need public services and public infrastructure that are, to paraphrase technologist Bruce Schneider, trusted to admit their flaws.[29]

PATTERNS

1. Systematically publish information about how systems change
2. Make data auditable
3. Publish rules as code

Systematically publish information about how systems change

Digital services and infrastructure are complex and change all the time. To enable public scrutiny, the teams operating them must systematically publish information about how systems work and how they are changing.

The diagram shows a common component used to implement 'single sign-on' to public services. The component is used by multiple services, including some high-volume and high-risk services such as social security, tax and voter registration. Recently, the team responsible for the component added a new feature enabling users to delegate access to third parties, such as accountants.

In advance of the changes taking effect, the team responsible for the common component published various bits of information. They added the new user needs they had identified for delegating access to an account to a public list of user needs. Example user journeys for delegating access were published to a design archive, including screenshots, support centre scripts and screen-reader recordings. The register of databases was updated to include information on the new data that would be collected to support the feature. Finally, when the feature was enabled, an explanation for the changes was published in the public version log.

A few weeks after launch there was an issue where users were unable to delegate access to their tax records because of a change to a screen where users manage their accounts. After the problem was remedied the team published an incident log detailing the issue and how it was fixed. Because the full list of services that use the component is linked to from the *register of services*, the incident report is automatically associated with those services.

Publishing information about how systems work and how they are changing means that lessons can be learnt. While

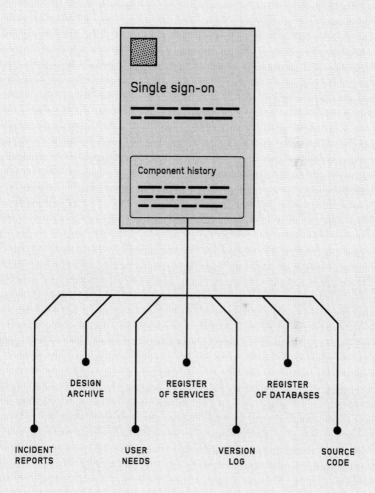

putting it on record might seem to be inviting criticism, it does the opposite: transparency removes much of the space for people to create their own narratives.

Make data auditable

When data is used to support multiple services, the opportunities for misuse grow. This is especially true when data is used to support services outside the public sector.

To identify and deter misuse and poor practice, there need to be ways of auditing how data is used. At the scale of an individual user, people should be able to check how data about them has been used. At the scale of the overall system, public auditors need to be able to check how services are using public data. Furthermore, they need to be able to find out if the custodians of data are maintaining its integrity and quality.

The diagram shows a register of land ownership that is used by a mix of public and private sector services. Three private sector services are shown: a commercial property selling service, a mortgage provider and an insurance service.

In this example a homeowner can inspect a journal (see chapter 3) to understand how data about their property has been used. If, for example, the mortgage provider had added something to their record, they would be able to see when and why this had occurred. If it looked as if it had happened in error, they would be able to raise the alarm.

The diagram also imagines a public 'data audit office' with responsibility for checking how services are using public data. This auditor can inspect a list of the services using the register of land and review their activities to ensure they are conforming with the rules. So, if the insurance company was found to be accessing data to train a machine learning model, for example, it might have its access suspended if this was outside the rules.

The auditor can also verify the integrity of the land ownership data. Specifically, it can check that records were created

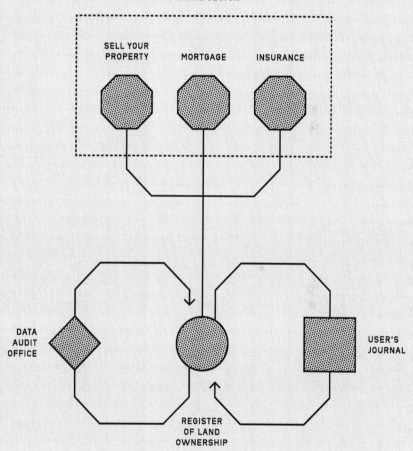

PRIVATE SECTOR

SELL YOUR PROPERTY

MORTGAGE

INSURANCE

DATA AUDIT OFFICE

USER'S JOURNAL

REGISTER OF LAND OWNERSHIP

by services with the correct permission; that once land has been registered, it remains registered; and that the database as a whole hasn't been tampered with.[30] So, if the register had been edited by a public official without permission, that would be identified.

Not only does making the use of data auditable by users and public auditors create the means to spot issues, it also creates an expectation that data will not be misused.

Publish rules as code

The rules that underpin services are the manifestation of policy and legislation. They need to be *inspectable* and *testable*, so society can verify that they are properly implementing the intent of the policy and legislation.

Helping society understand how digital systems work does not require special software or 'registers of algorithms': in most cases, it is enough to publish the code that services use to make decisions.

The diagram shows the rules for a public-facing service at its centre. If the service those rules support was a trade export service, they might encode the legislation on export certificates. If it was a welfare service, they might encode entitlement policies. For a property tax service, it might be the tax calculation.

There are four use-cases for the rules shown in the diagram.

1. The first is the service itself.

2. Next is parliamentary oversight. With appropriate technical support, parliamentarians can verify that the laws they pass are being adequately reflected in the rules.

3. Next, the diagram shows how testing the 'correctness' rules can be automated. Software tests are code that checks code, normally written by software engineers to make sure

PARLIAMENTARY
OVERSIGHT

PUBLIC-FACING
SERVICE

PUBLIC
AUTOMATED
TESTING

CIVIL
SOCIETY
RESEARCHERS

that their own code works properly. Public testing of rules could, for example, confirm that the rules used by a property tax service are calculating the correct amount of tax. This testing can happen automatically, with the results published online. One day, parliamentarians might even publish tests alongside legislation to confirm how they expect it to be implemented.

4. Finally, the diagram shows how civil society organizations are important users of rules. Like the parliamentarians, they would be interested in checking that the rules do not exclude members of the public from accessing their rights and entitlements.

As more high-stakes public services are digitized, transparency about how legislation and policy have been *implemented* must become an expectation, with virtually no exceptions.

STRATEGY

Automate transparency

At times, critiques about the ethics of digital systems seem to be willing digital services to be more like bridges and less like rivers. 'If only there was a more robust specification process.' 'If only "the algorithms" were subjected to more scrutiny during the procurement process.' (As if they are somehow discrete, purchasable things.) But these critiques are subject to the same category error perpetuated by software vendors to sell their wares by pretending they are static products. Modern digital practice means that there is no master document describing the system that is being built (and if there is, it's probably either out of date or a work of fiction demanded by a manager).

Digital services and infrastructure are complex, ever-changing socio-technical systems. Publishing information about how they work and when they change creates ground truth. This information can be used by parliamentarians, civil society groups and other experts to better understand them.

Prioritizing transparency is hard for organizations. Public servants may worry about being 'caught out' by some embar-rassing bug or other. However, that has to be weighed up against the very real harm that a lack of transparency can have.

Even if nothing goes wrong, opacity is no protection against negative headlines. There is a sort of 'digital fog' that forms around controversial services. In the absence of information, people start to assert that things work in a certain way because of either misunderstanding, personal politics or malice.

Digital practitioners working with organizations early on in their digital transformation find it's another thing to explain – something that often has to take its place in the queue behind user-centred design and agile practices. Smaller, less well-resourced digital teams may find it difficult to prioritize against work that has a more direct value to users. But for organizations of any size that are betting their future on digital transforma-tion, this is not something that can be dodged.

Fortunately, it doesn't have to be expensive. It is one of the paradoxes of digital that the act of explaining how things work becomes simultaneously harder and simpler.

It's harder for all the reasons set out in this chapter: the ongoing evolution of software, personalization, the opacity of data use and the rapid iteration of designs. These things are just inherently harder to explain and understand than a service delivered face to face with paper forms.

At the same time, they also become simpler because many of the things described above can be automated and the cost of publishing information online is effectively zero. The tools and practices that have grown up around software development can be repurposed, so transparency becomes an automatic byproduct of the creation of digital systems.

For larger organizations, things like version histories, design archives and a register of databases can be operated as components that are used by multiple services.

This is essentially what Apple and Google have done with their app stores. The app stores' public version histories explain new functionality. There is a (public) feedback mechanism in the form of ratings and comments. There are often some screenshots showing what a user can expect. Information about the organization operating the service is consistently presented, and the 'privacy labels' attempt to explain how data is used. It is telling that an iPhone app for identifying insects is probably more transparent than the average digital public service.[31]

Main points

1. Software is highly mutable.

2. Software evolves in response to its environment.

3. The work of digital teams is organized around change.

4. Change and personalization create an accountability gap.

5. Boundary objects help different groups to understand each other.

6. Many of the byproducts of digital design and development could act as boundary objects.

7. The public sector needs a largely true information environment about how services and infrastructure work.

8. Teams should systematically publish information about how their systems change.

9. The integrity of data and the use of data can be verified automatically.

10. Publishing rules as code means different groups can check if they are correct.

11. In the absence of transparency, people will make things up.

12. Transparency can be automated.

Examples: anatomyofpublicservices.com/examples/accountable

CHAPTER 10

Immunity to treachery

A rriving for work on the morning of the 23 September 1997, employees of Netscape found a giant icon dumped on the grass outside their office. The three-metre-high blue 'e' was the icon of Microsoft's Internet Explorer web browser and was a stage prop taken from a Microsoft launch event the previous evening, apparently by overenthusiastic Microsoft staff. On the prop rested a note: 'From the IE team.'

The prank was discovered by Netscape employees just before the local San Francisco press turned up. Netscape staff knocked over the stage prop and set their dinosaur mascot 'Mozilla' in the middle of the 'e' with a sign reading 'Netscape 72, Microsoft 18', referencing the companies' respective shares of the browser market.[12] The press dubbed it the 'Revenge of the nerds'.

In the mid 1990s, if you used a web browser at all, it would be Netscape Navigator. Its icon – a white letter 'N' rising like a moon over a dark horizon – was the gateway to the nascent World Wide Web.

In 1995 Netscape had a 90% market share, and it was poised to become more than just a web browser. Programmers were beginning to write plug-ins for Netscape using its newly launched API. Plug-ins written for Netscape would work across any operating system, so Microsoft's Windows risked becoming interchangeable with any other operating system. The direct relationship Microsoft had with its users would be lost, and some plug-ins might develop into direct competitors to its Office products too. Microsoft risked being 'disintermediated'. Investors had begun to notice, and the company needed a plan – not just to tackle Netscape, but for the internet in general.[3]

Microsoft's approach was set out in an internal email by Bill Gates in May 1995 titled 'The internet tidal wave'. To see off the risk of disintermediation, Microsoft would 'embrace' the internet and then 'extend' it.[4] The company developed its own browser, Internet Explorer, which it started pre-installing with Windows 95. Internet Explorer had its own plug-in system that it used to embed features of Windows and Microsoft's other software directly into web pages.

The plan worked, and by 1997 Microsoft had captured 76% of the browser market.[5] In defeat, Netscape went on to become the open-source Mozilla project. Its Firefox browser would eventually take back a significant chunk of the browser market and put web standards, rather than proprietary APIs, at the heart of the web. But before that, Microsoft's new market dominance led to a landmark antitrust case by the US government, which accused it of using Internet Explorer to create a monopoly. The government initially won the case, and the court ruled that the company would be broken in two, with the Windows operating system split off into a separate entity from the one responsible for applications such as Microsoft Office, but the decision to break up Microsoft was overturned on appeal.

In the final settlement between Microsoft and the US government, the former was mandated to share details of its APIs with third parties. The process was overseen by an independent panel of three people who would have full access to the source code to ensure nothing was being hidden.[6]

Rival software vendors also brought cases against Microsoft. Reporting on a case brought by Bristol Technology in 1999, the New York Times journalist Steve Lohr summed up their complaints:

Rivals have long said that Microsoft gives them only rudi-
mentary maps of the Windows APIs, making it impossible to
compete with the designs of Microsoft's own applications
programmers, who enjoy a detailed road map of Windows,
complete with little backroads and alleys that can be used
as shortcuts.[7]

The accusation was that Microsoft had an unfair advantage
because it tightly controlled the code that software developers
relied on. An internal email written by a Microsoft manager submitted as part of the Bristol Technology case seemed to confirm
this. It contained the line: 'To control the APIs is to control the
industry.' Microsoft and Bristol Technology went on to settle
their dispute, with Microsoft winning against antitrust claims
and Bristol Technology winning one case on unfair practices.[8]

These stories are recounted here because the first browser
war and the Microsoft antitrust cases demonstrated for the
first time, at least in any significant way, the potential for code
to set the rules that everyone building on top of that code had
to follow. These events demonstrated what legal scholar Law-
rence Lessig summed up in the adage 'code is law'. Software
code regulates behaviours, he said, so we must ask 'how the
code regulates, who the code writers are, and who controls the
code writers'.[9]

Following the court cases, people also started noting the
tendency of digital products to create powerful 'natural monopolies', and that, because software costs little or nothing to distribute, monopolization and disintermediation might happen
at speed.[10]

According to the social psychologist Dacher Keltner, power
is the ability to alter the states of others.[11] Digital services and
infrastructure can affect the distribution of power in society.
They can give and take power at the level of an individual system – the design of a particular service may use data to ensure
one group benefits more than another – but they can also create

zones of social and political contest between different groups by controlling who can and cannot operate services.[12]

And this is something that US justice department officials apparently understood during the US government's antitrust case. During the case one official told the journalist Steve Lohr: 'I think we should put a sign up in the war room here, "It's the APIs, stupid".'[13]

The use of digital in the public sector gets framed as 'innovation', with the connotation that innovation is something brought into the public sector by the private sector. One of the reasons this is objectionable is that the application of technology opens up new contested zones between the public and private sectors. As public services become digital public services, governments could find themselves at risk of disintermediation, just as Microsoft feared it was from Netscape in 1994.

During the Covid-19 pandemic Apple and Google created a contact-tracing API for their mobile operating systems.[14] The approach they implemented meant there was no central database of contacts and all the data was anonymous. They also set a 'one app per country' limit and prevented location data from being recorded.[15] The one-app limit removed the ability to meet different types of need (e.g. an app for healthcare workers that recorded if they were working on a non-Covid ward).[16] Some epidemiologists also argued that a centralized model would lead to better outcomes.

This led to many countries trying to get Apple and Google to change their approach.[17] Nonetheless, most countries created their contact-tracing app based on the Apple and Google API. As researcher Jonathan Albright (who compiled data about the technology used in different governments' Covid-19 responses) put it, health agencies around the world 'collapsed their efforts' into app marketplaces and APIs built by two companies.[18]

When the UK government tried to create a centralized contact-tracing app for England and Wales, they ultimately failed

– in part, it seemed, because the constraints that Apple and Google had set made the project harder.[19]

To see this as a case of the public sector not understanding technology would be unfair. Contact tracing was part of a massive effort by public sector technologists around the world to create new services for the Covid-19 response. In addition to contact tracing, these included services related to testing, vaccination booking, financial support and public awareness. The UK government would eventually go on to create a successful contact-tracing app using the Apple and Google API, which is estimated to have saved 9,600 lives in its first year of use (although how a centralized model might have performed is destined to remain a counterfactual).[20]

The dominance of Apple and Google's APIs had some other interesting effects too. It became hard to understand where the government parts of the system ended and the Apple and Google parts started. In September 2020 the journalist Rowland Manthorpe reported that, in England and Wales, some of the Covid-19 exposure notifications people were receiving were actually default messages from Apple and Google, not from the UK government.[21] It's doubtful that the people who received them understood that, though. Furthermore, while there were calls for governments to publish the source code of their apps, there were few calls for greater transparency regarding the parts operated by Apple and Google.

It's easy to see why Apple and Google took the approach they did. Governments can and do abuse their power. iOS and Android are global platforms, and their API was destined to be used in a wide range of political contexts. Both companies are American, so it would be surprising if they were not also unconsciously calibrating their response to the risks of their own country's political and healthcare systems.

Regardless of the rights and wrongs of the centralized versus decentralized contact-tracing approach, the intervention by Apple and Google represented a form of policymaking by API.

The debate about efficacy had to play in the space demarcated by the API.

Another example of a contest between the public and private sector is seen in the US tax system. Unlike countries that have automated the process, the United States still requires taxpayers to file a tax return every year. Most people use commercial software to do this. Having secured their dominant position, the software providers now actively lobby against the public sector offering 'competing' services. In 2019 ProPublica reported on a bill before congress that would have made the arrangement permanent, effectively banning the US Internal Revenue Service from ever providing a service to complete a tax return. The suggestion is that this is due to lobbying from private tax-tech firms.[22]

Since the early 2000s US tax software companies have provided a free option, called 'Free File', for low-income taxpayers to complete their returns. In exchange for providing this service, the government agreed not to create its own system, but another ProPublica investigation accused the makers of the widely used TurboTax software of hiding the free version of its software from search engines and of using various methods to drive people towards the paid versions.[23]

Ultimately, the relevant clauses were removed from the bill and the Biden administration would go on to launch a US government service called DirectFile as a pilot in 2023. Following a successful trial, DirectFile was made permanent in 2024, but as of June 2024, the Republican-controlled House of Representatives was trying to get the service's funding removed.[24]

As governments create more digital infrastructure, more opportunities for this sort of behaviour are likely to be created. As detailed in chapter 2, opening up rules as APIs and data infrastructure to private sector services is a good strategy for reducing the administrative burden of users, but making it easier for the private sector to create services could also make disintermediation easier and more common.

Governments will need to act as Microsoft was accused of acting to ensure that they are not disintermediated. They will need to take an opinionated stance on what services are allowed to operate and how those services interact with digital public infrastructure. They will also have to ensure they maintain control over foundational data, rules APIs and the ability to issue credentials (see the 'Design against disintermediation' pattern later in this chapter).

Digital infrastructure doesn't create zones of contest only between the public and private sectors: it creates them *within* the public sector as well, and it represents a new version of an old political battle.

Poplarism, named after the Poplar district of London, was an early twentieth century movement that showed the power of local government to challenge unfair central government policies.

The system of property taxes in 1920s London was organized such that Poplar, as a poor borough, had to set higher rates than much richer areas to pay for London-wide policing, water services and the 'Poor Laws' (a system of poor relief and 'workhouses', with its origins in the 1500s). Local councillors, who managed the collection of taxes, refused to pass them to central government, instead spending them on local priorities.[25] The councillors were sent to Holloway and Brixton prisons, but the ensuing campaign resulted in changes to the tax system and led to the end of the Poor Laws.*

As with tax collection, common digital infrastructure represents a 'single point of control'. The public institutions operating that infrastructure have the potential to exert control over all the public services that rely on that infrastructure, including those provided by another tier of government. So, a single component for authenticating users operated by central government

---

* Shortly thereafter, central government also passed laws to make sure Poplar and other boroughs could never repeat the trick.

represents a potential gatekeeper to services, shutting users out, as much as it does a potential enabler of access to those services. Common infrastructure may also have a 'line of sight' over a wide range of interactions across the public sector and beyond. A single appointment-booking platform might contain information about everything from healthcare to education and housing. A single notifications system might process messages about everything from registering to vote to booking a vaccination to applying for a visa.

Historically, the public sector has been organized around a mix of location, policy and avoiding concentrations of power. Government agencies have clear policy briefs (the environment, benefits, tax, driving, and so on). Local, regional, and central government have defined geographic jurisdictions and responsibilities. Auditors, courts and commissions exist to hold other parts of the system to account.

Rather than being hierarchically and geographically organized, the future could be inherently more networked.

Services designed around users, not the structure of government, may cut across geography and hierarchy. Credentials from one service might be the input into many other services. Rules exposed as APIs could mean different users getting the same task done in very different ways, in services operated by different organizations. Common components and data infrastructure mean organizations will be dependent on infrastructure operated by other organizations to deliver their core mission.

Together, these things represent a reorganization of the work of government around a network of rules APIs, components and canonical datasets, so that public servants, businesses and others can deliver radically better services to the public, more safely, efficiently and accountably.

Reorganizing the work of government doesn't just create new dependencies between parts of the public sector, either: some organizations will stop doing things they have long done, and others may start doing new things.

If central and local government have conflicting opinions about the delivery of a service, common infrastructure may create opportunities to restrict other activities. As with the Covid-19 exposure notifications APIs, common infrastructure created by one organization may limit the space that the policies of another can operate in.*

Organizations that used to operate public-facing services may find that they no longer have a direct relationship with their users. When Argentina created its digital driving licence, for example, the agency involved just provided an API that was used by the miArgentina platform, which is operated by the country's central digital agency.[26] Similarly, the rules APIs of the US Department of Veterans Affairs are used by non-governmental veterans' organizations to submit claims on behalf of those they are supporting.[27]

Rather than maintain their own copies of data, organizations operating public services will find themselves relying on common data. For example, the National Resident Population Register is a central population register for Italy that replaces regionally held registers.[28] In India, registers of voters are maintained by regional Electoral Registration Officers who are constitutionally independent, but a data-cleansing exercise in 2022 used code written by the Election Commission of India.[29]

Efficiencies of scale and the need to maintain appropriately skilled teams to operate infrastructure mean that natural monopolies for digital infrastructure feel as inevitable in the public sector as they do in the private sector. The centralization of digital infrastructure does not, as digital anthropologist Payal Arora points out, have to mean control, but it does raise questions about what constitutes democracy.[30]

The institutions operating shared infrastructure have the potential for great power and need the capacity for responsibility. They need to be constituted to actively defend the data

---

* The opportunities for neo-Poplarism in a world of centrally maintained infrastructure seem minimal.

they hold about users against misuse, they need to have enough institutional clarity, and they need the robustness to act in the public interest.

It has become common in the digital sector to equate trust with privacy and transparency – something to be solved with technical measures or better design. But technically mediated privacy and transparency are better seen, respectively, as safety and accountability measures.

Trust in public services is a matter of democracy, not technology or design. It is a question of if the public trust an institution to deliver the 'public good' they are tasked with, or if they believe it is acting counter to that public good.[31] The National Health Service in the United Kingdom is one of the country's most trusted public institutions. But public trust does not extend to its data being used for research and 'innovation' by the private sector. That's because it challenges the idea of an institution tasked solely with the delivery of a public good, i.e. providing healthcare, from which the National Health Service earns its trust.[32] You can't solve that sort of contradiction with design thinking or technical architecture.

The creation of the of next-generation public services and infrastructure will require institutions with clarity of purpose and clarity of accountability. The creation of GOV.UK was accompanied by the creation of the Government Digital Service: a dedicated organization at the centre of government. Many of India's common components have also been created within new institutions. DigiYatra is a common component used by airports and airlines to replace boarding passes with facial recognition. It is operated by the DigiYatra Foundation, in which the government has a 26% stake via the Airports Authority of India with the rest being divided among major airports. The Open Network for Digital Commerce – a public data exchange for implementing e-commerce transactions that could one day rival Amazon – is a private non-profit company established by the Department for Promotion of Industry and Internal Trade. The Goods and Services Tax Network, which operates rules APIs for the management of sales taxes by states, is a non-profit company owned by Indian national and state governments and private sector financial institutions.[33] Aadhaar, the digital identity system, is operated by the Unique Identification Authority of India. Created by an act of parliament, it is forbidden from sharing the underlying biometric data linked to an Aadhaar number.[34]

While these new institutions have some clarity around their missions and accountability, there is much that remains contested. Civil rights lawyers have raised concerns about the lack of oversight and transparency over how data is handled in DigiYatra, for example.[35] What can and cannot be done with Aadhaar is an area of ongoing legal action, debate and activism too: the Calcutta High Court has, for example, ruled that the Ministry of Home Affairs could not deactivate the Aadhaar identities of a family who were suspected of being Bangladeshi citizens.[36] Proposals to allow private sector use of Aadhaar led to activists from the Rethink Aadhaar campaign arguing that it crossed a red line set in the 2018 Supreme Court judgement that limited such use.[37]

Rather than being a signal that India has got it wrong with its approach to digital infrastructure, we should see it as a result of having got there first. The inherent mutability of digital systems and the opportunity for 'mission creep' that that mutability affords mean that legal and civil society activity should be considered as something that is necessary for any democracy creating digital infrastructure and services.

There are examples of institutions from outside of India too. In the United Kingdom, the success of the Government Digital Service was in part down to the fact that it was a new organization with responsibilities across government. Universal Credit was a new organization too. It may have been housed in an existing government department, but in terms of everything else it was a new institution.

The Nordic Institute for Interoperability Solutions is a collaboration initially formed by Finland and Estonia. Established in

2017 its main purpose is to develop and manage the X-Road data infrastructure that is used by both countries.

Back in the United Kingdom again, the LocalGov Drupal software used by local governments to build and operate their websites is operated as a cooperative, created through seed funding from central government. Local governments who use the software and vendors who customize it can pay to be members of the cooperative, giving them voting rights over how the product develops.[38]

The Australian federal government's plans for digital identity imagine it being operated by an oversight authority (reappointed every five years) that cannot be 'directed'. This is in order to give it some distance from ministerial control.[39]

These new institutions all represent the weaving of digital systems into the democratic fabric of a country. They are treating digital not as something outside the purview of the public sector, but as something core to the operation of a democracy.

There are echoes of something earlier here too.[40] The Metropolitan Water Board was set up in 1903 to take over the operation of the water system in London from an ad hoc, unsafe and unaccountable set of private companies. Here is what the government commission that led to the water board's creation recommended:

> The water supply should be brought periodically and automatically under the observation of Parliament, and that it should include delegates of the Local Government Board, in order that the influence of the executive government should be continuously felt.[41]

This was the same water board whose funding the Poplar councillors would, a few years later, stage their protests against. They were only able to effect change in the way they did because water management had become a public institution. Public ownership is not only a question of efficiency, it enables public debate.

Similarly, in 1905 the Fabian Society* was arguing for greater public control of the new electrical power networks and warning of the risks of the private ownership of critical public infrastructure:

> If gigantic and tyrannical trust monopolizing the production and use of electric power are not to dominate our children as the railway companies dominate us, we must see that the community secures at the outset effective and systematic control over the new force.[42]

First local government and then national government would go on to create a publicly owned system of power generation and distribution in Britain.

There are echoes further back too. In 1513 the Corporation of Trinity House of Deptford Strond was created when mariners on the River Thames were so concerned about the poor conduct of pilots on the river that they asked the king for a licence to

> improve the art and science of mariners; to examine into the qualifications, and regulate the conduct of those who take upon them the charge of conducting ships; to preserve good order, and (when desired) to compose differences in marine affairs, and, in general, to consult the conservation, good estate, wholesome government, maintenance and increase of navigation and sea-faring men.[43]

Today, and with its remit expanded over the centuries, the same corporation manages the lighthouses, maritime communication systems and buoys in the coastal waters of England and Wales. Its shared infrastructure supports not only commercial operations but also the business of government.

---

* The Fabian Society was founded in 1884 with the aim of promoting gradual, rather than revolutionary, reform to secure social and economic improvements in Britain.

Trinity House is not part of government, per se: it's a corporation created by 'royal charter'. It is paid for through a mix of fees (called 'light dues') paid by commercial vessels, and an endowment and oversight is provided by a court of thirty-one people, drawn from a wider group of laypeople with marine experience.

The digital age needs new institutions to deliver and support public services. As with Trinity House, the path dependency imposed by existing systems and institutions will mean it will be different in each nation state, so it is not a case of 'doing an Estonia' or 'doing an India'. The exact constitution of and interdependency between the different parts is not something that can be 'designed': it will be a complex ecosystem. Society as a whole will need to be as vigilant against creating concentrations of power as they are in demanding better services.

There is a concept in evolutionary biology called the 'evolutionarily stable strategy' that serves as a principle worth aspiring to here. An evolutionarily stable strategy is one that, 'if most members of a population adopt it, cannot be bettered by an alternative strategy'. The standard example – given by Richard Dawkins in his book *The Selfish Gene* and based on the work of the geneticist and evolutionary biologist John Maynard Smith – is a game between participants in a hypothetical ecosystem of aggressive 'hawks' and non-aggressive 'doves'.

In a population that is entirely hawks, a mutation that creates a dove will lead to a huge advantage for the dove genes because, although it will always lose by running away, it will never get hurt and the genes will spread throughout the population. In a population that is mostly doves, a hawk has a huge advantage, since it will never meet another hawk and risk injury. Rather than oscillating wildly between the two strategies, a population will arrive at a stable ratio where neither of the strategies beats the other. It's not that it is good for the individuals, it's that it is, as Dawkins put it, 'immune to treachery from within'.

That is what a networked public sector should aspire to: immunity to treachery.

PATTERNS

1. Make accountability clear
2. Design against disintermediation
3. Value and governance change with scale

Make accountability clear

Services and infrastructure point in opposite directions. The incentives to run a great public-facing service rarely align with those needed to operate great infrastructure. For services to meet the needs of the public, they need clear ownership and accountability. And if components and data are to meet the needs of multiple services, they need clear ownership and accountability too.

The diagram shows a series of services supported by common components and data. It describes who operates them and who they are accountable to.[44]

The services and the associated rules and credentials are operated by 'service managers'. A service manager's job is to prioritize the needs of the users of the service and implement the policy intent of government ministers with responsibility for a specific policy area.

Components are operated by 'platform directors'. Like service managers, their job is to understand the needs of their users, but the primary users of components are the teams creating services, not the public. Platform directors are accountable to government ministers with a cross-government brief.*

Data is managed by 'custodians' or 'registrars'. Their role is to maintain the quality and integrity of the data, ensure that standard identifiers are used, and understand the needs of the services that use the data. They may be responsible either to a minister, directly to parliament, or to another elected body.

However, clear ownership and accountability are nothing without a culture that prioritizes the health of the overall ecosystem. Public sector leaders need to be comfortable allocating time and money both to support work that could have a wider benefit and to help everyone feel it is part of their job to work together.

---

* If a component for making payments to users is operated by the minister with responsibility for transport or farming, then it is unlikely they will prioritize the full range of potential users.

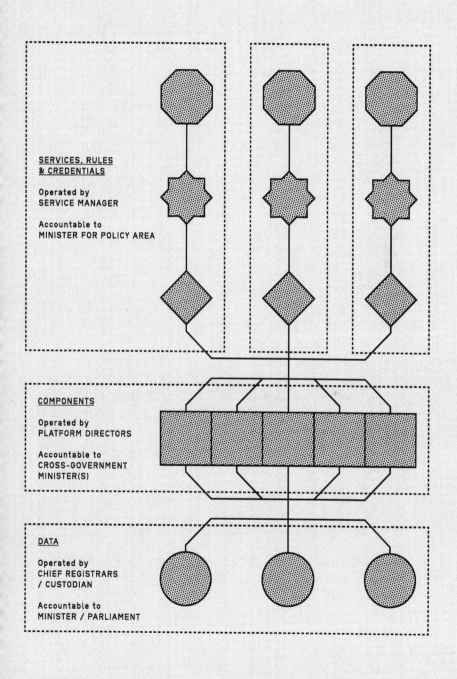

SERVICES, RULES & CREDENTIALS

Operated by
SERVICE MANAGER

Accountable to
MINISTER FOR POLICY AREA

COMPONENTS

Operated by
PLATFORM DIRECTORS

Accountable to
CROSS-GOVERNMENT
MINISTER(S)

DATA

Operated by
CHIEF REGISTRARS
/ CUSTODIAN

Accountable to
MINISTER / PARLIAMENT

Design against disintermediation

There are needs that government will never meet. Allowing public and private sector services to be built on top of common rules means that more needs can be met (see chapter 2), but the underlying infrastructure must be operated in a way that minimizes the risk of disintermediation by those private sector services.

Disintermediation in digital public services occurs when a commercial service provider replaces or exerts control over the data rules and credentials critical to the running of the public service. Public services can defend against this risk by ensuring they maintain control over rules, credentials and data.

The diagram shows a government service and a complementary private sector service. Both allow a user to file a tax return. The commercial service makes use of the credentials, rules and data operated by the government rather than maintaining its own. The private service has a direct relationship with users, but it cannot use that to take ownership of the underlying rules and data. It also cannot 'mint' its own credentials.

The consequences for commercial services who attempt to misuse their position must be clear, including the situations in which they can have their access limited.

The privatization of the next generation of public services may be harder to spot. There needs to be a clear understanding and ongoing monitoring of the risks by public sector leaders.

Value and governance change with scale

Components and data, operated as public goods, can create huge amounts of public value. But with that value, they can also generate political power.

An identity verification component that only supports central government services has less opportunity to generate public value than one used for everything from applying for welfare,

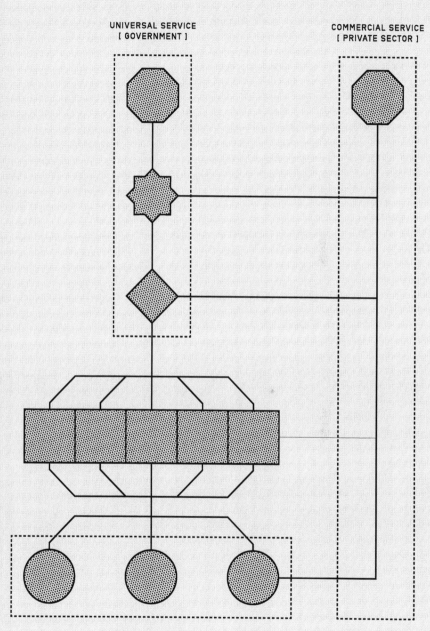

UNIVERSAL SERVICE
[GOVERNMENT]

COMMERCIAL SERVICE
[PRIVATE SECTOR]

CANONICAL DATA
[GOVERNMENT]

to buying a parking permit, to opening a bank account. But an identity verification component operated by a central government agency but used by regional government and the private sector puts enormous power in the hands of the central government. It could, for example, limit access to identity verification for local government services that it disagreed with politically.

For these two reasons, governance becomes more important the broader is the use of components and data. As the diagram illustrates, this common infrastructure can be used by

- a single tier of government – used by multiple departments or agencies within a government (central, state or municipal);

- the whole of the public sector – used across central government, regional government and city government; and

- society wide – used by the public sector, the private sector and the third sector.

Components and data that have wide usage need a clear institutional home that is protected from the day-to-day whims of political institutions. They also require a legal framework that secures the rights of the public and guards against concentrations of unaccountable power.

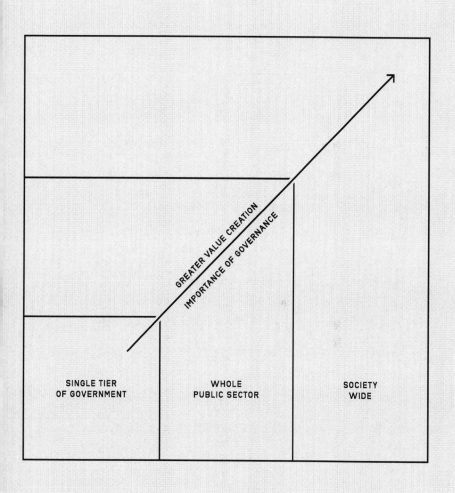

GREATER VALUE CREATION

IMPORTANCE OF GOVERNANCE

SINGLE TIER
OF GOVERNMENT

WHOLE
PUBLIC SECTOR

SOCIETY
WIDE

STRATEGY

Grow new institutions

The experience of India, the United Kingdom, Estonia and others has shown that digital-age public services and infrastructure require new organizations to operate them.

The need for new institutions is partly about addressing what Jennifer Pahlka (founder of Code for America) calls the 'implementation crisis', created by governments without the capability to deliver functional digital services.[45] New institutions can fill the capability gap. But the types of organization described in this chapter also represent a reorganizing of the work of government for the digital age so that it is clear what 'public good' they are tasked with. If a government genuinely needs to operate a digital identity system, an e-commerce network or an automated healthcare system, then it needs to be very very clear whose task it is, what value will be derived and for whom. Without that, there can be no trust, however well designed public services are considered to be by professional standards.

Luckily, like good digital services and infrastructure, the teams creating them can start small and grow too. The Government Digital Service was itself a 'working prototype'.[46] Aadhaar was a 'start-up' in government and, reflecting on its progress, two of its founders proposed that 'a team of 100 carefully selected individuals can fix all the major problems that ail India'. That is a bold claim, but the principle of starting small and growing a team into an institution is the model that governments will need to follow.

Main points

1. Disintermediation occurs when one service is built on top of another, severing the direct relationship with users.

2. Controlling APIs is a means to limit the risk of disintermediation.

3. Digital services and infrastructure create zones of political contest between the public and private sectors.

4. Commercial APIs that the public sector relies on can limit what the public sector can do.

5. In the private sector, controlling APIs can be anti-competitive; in the public sector, it can protect against capture.

6. Digital services and infrastructure also create zones of political contest *within* the public sector.

7. Government agencies no longer have a direct relationship with their users, and they become reliant on other organizations for data and components that are core to their mission.

8. Institutions operating shared infrastructure could become very powerful.

9. Trust in public sector institutions derives from their ability to deliver a public good.

10. New infrastructure needs new institutions.

11. Civil society scrutiny is a precondition for creating safe digital infrastructure and services.

12. The exact form of the public sector ecosystem is not something that can be designed up front; instead, the guiding principle should be 'immunity to treachery'.

Examples: anatomyofpublicservices.com/examples/immunity

Epilogue

I f code is law, then software – the manifestation of code in the world – is politics.

Ultimately, Britain's 1960s 'white heat of technology' revolution that created Girobank would fail on its own terms. Technology was supposed to be a meritocratic force within the public sector, but the civil service organized its technology around its gender- and class-based hierarchies. It was that failing combined with government consolidation of the nascent British computer industry (a lineage that included Leo, the world's first business computer) into International Computers Limited (ICL) that would see it become reliant on outsourced technology companies.[1]

In under a century the British Post Office went from *the* primordial digital institution to one whose outsourced IT systems have become the archetype for unaccountable failed technology. Bugs in its Horizon computer system led to Post Office staff being wrongly convicted of stealing.[2] Horizon was created by the outsourcing company Fujitsu, which had absorbed ICL.

The post-war digitization of the United Kingdom's tax and benefits system, the creation of automated institutions such as Girobank, and the decision to teach children to program with the BBC Micro were all bets on the future. Those kinds of bet have been long absent from British politics, but looking to the energy behind digital infrastructure in Estonia and India, it's hard not to see the similarity.

In the introduction, I asked to what extent is the digitization of the public sector inevitable? Hopefully, thinking about that question helped you calibrate the relative importance of the various approaches. Maybe reading this book will have changed some minds, making you think it is more or less likely than you did before. I'd now like to propose two complementary questions.

1. What should the bounds of the public sector be in the digital age?

Perhaps the biggest decision that governments have to make is which organizations they choose to create. It defines the areas in which governments claim a right to operate. I don't mean rearranging the existing 'machinery of government': I mean new organizations operating in new areas.

That was what the UK government did when it created Girobank. It claimed a right to operate in the financial sector. Similarly, the Indian government has decided to directly compete with commercial platform providers such as Amazon and Google in the name of digital sovereignty. While in France, Le.Taxi is the French government's answer to Uber.[3] The exact boundaries will be different in each society.

2. What will 'the politics' of digital services and infrastructure be?

Public policy has its own version of the law of conservation of energy. Nothing is new; issues can only be transferred and subdivided, but never destroyed. In the United Kingdom, house building, basic bank accounts, railway ownership, means testing and property taxation have been permanent political fixtures over the past century.

Digital, in the public sector, is not solely a question of capability – solving problems in the same way using new tools. Digital services and infrastructure are 'artefacts with politics', as the political theorist Langdon Winner put it.[4] They change how public policy issues are addressed and influence how power is distributed. For example, 'tax bands' are an analogue invention, but digital allows property taxes to be calculated on a continuum based on income or distance. Eligibility checking can be semi-automated, so hyper-means-testing becomes easier.

Disentangling the political from the technical can be tricky. China's Social Credit, built on the principle of 'trust-breaking here, restrictions everywhere', shows how access to services can be cut off abruptly.[5] But so does the EU-wide digital driving licence that enables EU-wide driving bans for dangerous driving.[6] It is likely that the true politics will only emerge through use.

*

Both of these questions require the people who create our big-P politics to understand the affordances of digital. A failure to do so is to cede what are political decisions to those implementing systems. Software is politics now.

About the author

Richard Pope was part of the founding team at the UK Government Digital Service. He created many of the design concepts for GOV.UK as its first product manager, and he co-authored the Government Service Standard, which all UK government digital services must meet. GOV.UK won the Design of the Year award in 2013 and is used by millions of people every day. The Government Service Standard has been copied by governments around the world.

As a civil servant, he went on to work with teams across the UK government to redesign government services in policy areas as diverse as employment, land registration and welfare. As part of his work on welfare, he was responsible for the designs of the digital account for Universal Credit, the United Kingdom's single social security service. Its use of an online journal and task list was unique in a digital public service. They brought much-needed clarity to an incredibly complex public policy while placing a greater obligation on government to explain its demands and actions.

In 2019, after leaving the Civil Service, he published the report 'Universal Credit: digital welfare', which called for government to share more of the benefits of digitization with the public and to embed transparency and accountability into the development of its services.

Before joining the Civil Service he co-founded Rewired State: a series of hack days that aimed to get more developers and designers working on issues of digital government. Many of the participants went on to work in government.

In 2008 he set up the digital team at Consumer Focus (the United Kingdom's statutory consumer rights body), an early example of a digital team in a public sector body.

He has also been a volunteer with MySociety.org; campaigned to open up UK postcode data; and created the election monitoring website electionleaflets.org, which generated dozens of news stories during the 2010 UK general election, and the PlanningAlerts service, which combined urban planning notices from hundreds of local governments into a single, simple service.

In addition to the public sector, he has also worked in several digital start-ups, including moo.com.

In 2018/2019 he was a senior fellow at the digital HKS initiative at the Harvard Kennedy School, focused on the subject of Government as a Platform, and he has lectured on the subject at the University of Cambridge, University College London and IE University.

He lives in London.

Notes

Preface

1 Edward Scott. 2020. Family Allowances Act 1945. House of Lords Library, 11 June (https://lordslibrary.parliament.uk/family-allowances-act-1945/).

2 British Movietone. 1968. Giro Banking. YouTube video, 19 August (www.youtube.com/watch?v=4C3WBm6pgzU). Glyn Davies. 1973. *National Giro: Modern Money Transfer*. George Allen & Unwin.

3 Marie Hicks. 2018. *Programmed Inequality*. MIT Press.

4 Owen Hatherley. 2021. The signs that make a city. *Tribune*, 15 April (https://tribunemag.co.uk/2021/04/the-signs-that-make-a-city).

Introduction

1 Bruce Schneier. 2023. AI to aid democracy. *Schneier on Security*, 26 April (www.schneier.com/blog/archives/2023/04/ai-to-aid-democracy.html).

2 Douglas Coupland. 2014. *Kitten Clone*. Random House Canada.

3 It's also inspired by Anthony Sampson's books on the 'anatomy' of Britain.

4 Jonathan Meades (writer and presenter). 1997. Remember the future. Episode of *Even Further Abroad*, BBC (www.youtube.com/watch?v=WWRmBXhC7Cs).

5 Sherry Turkle. 2005. *The Second Self: Computers and the Human Spirit*, 20th anniversary edition. MIT Press. Sherry Turkle. 2021. *The Empathy Diaries: A Memoir*. Penguin.

6 Tim O'Reilly. 2010. Government as a platform. In *Open Government*, chapter 2. O'Reilly (www.oreilly.com/library/view/open-government/9781449381936/ch02.html).

Chapter 1. Burden eliminated

1 Stephanie Ward. 2016. *Unemployment and the State in Britain: The Means Test and Protest in 1930s South Wales and North-East England*. Manchester University Press. Nicholas Timmins. 2023.

Why has the UK's social security system become so means-tested? IFS Deaton Review, 1 February (https://ifs.org.uk/inequality/why-has-the-uks-social-security-system-become-so-means-tested/).

2   Andrew Thorpe. 2001. *A History of the British Labour Party*. Palgrave Macmillan. Tony Benn in various speeches at The Left Field, Glastonbury Festival. Nicholas Timmins. 2017. *The Five Giants: A Biography of the Welfare State*. HarperCollins UK.

3   William Beveridge. 1942. *Review of Social Insurance and Allied Services*. His Majesty's Stationary Office.

4   For example: Social Security Administration. 2023. Electronic W-2 filing user handbook. December (www.ssa.gov/employer/bsohbnew.htm).

5   Donavon Johnson and Alexander Kroll. 2023. The effect of electronic program applications amidst the politics of administrative burden. *Government Information Quarterly* 40(2): paper 101808 (https://doi.org/10.1016/j.giq.2023.101808).

6   Alex Clegg, Deven Ghelani, Zoe Charlesworth and Tylor-Maria Johnson. 2023. Missing out: £19 billion of support goes unclaimed each year. Policy in Practice, April (https://policyinpractice.co.uk/wp-content/uploads/Missing-out-19-billion-of-support.pdf).

7   Charlie Moloney. 2023. Homeowners left out of pocket after two-year delays at UK Land Registry. *The Observer*, 27 August (www.theguardian.com/money/2023/aug/27/homeowners-left-out-of-pocket-after-two-year-delays-at-uk-land-registry). 'HM Land Registry forms': www.gov.uk/government/collections/hm-land-registry-forms#land-registration-forms.

8   NatCen Social Research and the Centre for Sustainable Energy. 2015. Data sharing to target fuel poverty. Report, August (updated December) (https://bit.ly/4cEqaS7).

9   richardjpope, twitter.com 'What administrative burden looks like' (https://x.com/richardjpope/status/1225370015408492546).

10   Richard Pope. 2011. A few design rules for Alpha.gov.uk. Government Digital Service blog, 28 April (https://gds.blog.gov.uk/2011/04/28/alpha-gov-uk-design-rules/).

11   Richard Pope, 'Moo.com UX Rules': https://principles.design/examples/moo-com-design-principles.

12   'The Trembita system, supported by the EU, facilitates digital governance development in Ukraine': https://decentralization.ua/en/news/13347. Diia. 2022. EAid: government payments in Diia app. YouTube, 24 May (www.youtube.com/watch?v=la-W8t4ZSKM).

13   Sneha Kulkarni. 2022. Soon, changes made to Aadhaar details will auto-update key DigiLocker documents. *Economic Times*, 16 March

(https://economictimes.indiatimes.com/wealth/personal-finance-news/soon-changes-made-to-aadhaar-details-will-auto-update-key-digilocker-documents/articleshow/90260409.cms).

14 Justin Petrone. 2023. No more digital garbage: real-time economy to streamline transactions and services. E-Estonia, 15 March (https://e-estonia.com/no-more-digital-garbage-real-time-economy-to-streamline-transactions-and-services/).

15 Cabinet Office. 2010. The coalition: our programme for government. Policy paper, May (www.gov.uk/government/publications/the-coalition-our-programme-for-government). Iain Duncan Smith. 2010. Our contract with the country for 21st century welfare. Speech to the Conservative Party conference, 5 October (https://bit.ly/3S8hg7a). Department for Work and Pensions. 2010. Universal Credit: welfare that works. White Paper, November (https://assets.publishing.service.gov.uk/government/uploads/system/uploads/attachment_data/file/48897/universal-credit-full-document.pdf).

16 Richard Pope. 2020. Support finding work. In Universal Credit: Digital Welfare. Part Two Digital (https://digitalwelfare.report/support-finding-work).

17 Richard Pope. 2020. Responsibility and complexity. In Universal Credit (https://digitalwelfare.report/responsibility-and-complexity).

18 Richard Pope. 2020. Annex 1: change of circumstances. In Universal Credit (https://digitalwelfare.report/annex-1).

19 Derek du Preez. 2013. DWP finally admits Universal Credit IT not up to scratch – calling in GDS. Computerworld, 10 July. (The original source is no longer online but is linked to from www.rightsnet.org.uk/forums/viewthread/5040/#20651.) Rightsnet forum, 'Discussion: Universal Credit project is officially "reset" / changing direction / will start again almost from scratch / technology scrapped …': www.rightsnet.org.uk/forums/member/3221/viewthread/5089/#21709.

20 Glick Bryan. 2014. Introducing the Devereux–Hodge shambolicness scale for rating progress of Universal Credit. Computer Weekly, 16 December (www.computerweekly.com/blog/Computer-Weekly-Editors-Blog/Introducing-the-Devereux-Hodge-shambolicness-scale-for-rating-progress-of-Universal-Credit). National Audit Office/Department for Work and Pensions. 2014. Universal Credit: progress update. Report, 26 November (www.parliament.uk/globalassets/documents/commons-committees/public-accounts/Universal-Credit-progress-update.pdf).

21 National Audit Office. 2014. Universal Credit: progress update. Report, 26 November (www.nao.org.uk/reports/universal-credit-progress-update-2).

22 Design Museum, 'Beazley designs of the year overview': https://
designmuseum.org/design/beazley-designs-of-the-year-overview.
Cabinet Office, Government Digital Service and Rt Hon. Nick Hurd.
2013. GOV.UK wins design of the year 2013. Press release, 16 April
(www.gov.uk/government/news/govuk-wins-design-of-the-year-2013).
Rosie Taylor. 2013. And the award goes to boring.com! Government
website beats off 100 others to be named world's best design. *Daily
Mail*, 16 April (www.dailymail.co.uk/news/article-2310191/And-award-
goes-boring-com-Government-website-beats-100-named-worlds-
best-design.html).

23 Government Digital Service. 2012. Government design
principles. Guidelines, 3 April (www.gov.uk/guidance/
government-design-principles).

24 TheyWorkForYou, 'Electronic government': www.theyworkforyou.
com/wrans/?id=2014-01-13e.181777.h&s=%22user+needs%22#g181777.
r0.

25 Tudor Rickards. 2006. Enid Mumford. *The Guardian*, 3 May (www.
theguardian.com/news/2006/may/03/guardianobituaries.obituaries).
David Avison, Niels Bjørn-Andersen, Elayne Coakes, Gordon B.
Davis, Michael J. Earl, Amany Elbanna, Guy Fitzgerald, *et al.* 2006.
Enid Mumford: a tribute. *Information Systems Journal* 16(4): 343–82
(https://doi.org/10.1111/j.1365-2575.2006.00225.x).

26 Enid Mumford. 1983. Participative systems design: practice and
theory. *Journal of Occupational Behaviour* 4(1): 47–57 (www.jstor.
org/stable/3000226). Enid Mumford and Harold Sackman (eds). 1975.
*Human Choice and Computers: Proceedings of the IFIP Conference
on Human Choice and Computers, Vienna, April 1–5, 1974.* North-
Holland Publishing Company/Elsevier. Enid Mumford. 2000. Socio-
technical design: an unfulfilled promise or a future opportunity? In
Organizational and Social Perspectives on Information Technology,
pp. 33–46. Springer (https://doi.org/10.1007/978-0-387-35505-4_3).

27 Rob Kling. 1977. The organizational context of user-centered
software designs. *MIS Quarterly* 1(4): 41–52 (https://doi.
org/10.2307/249021).

28 Mumford. 1983. Participative systems design.

29 Department for Work and Pensions. 2018. Universal Credit
programme full business case summary. Report, June (https://
assets.publishing.service.gov.uk/media/63456b85d3bf7f618c35e4d2/
uc-business-case-summary.pdf).

30 Naeha Rashid. 2020. Deploying the once-only policy: a privacy-
enhancing guide for policymakers and civil society actors. Policy
Briefs Series, Ash Center for Democratic Governance and Innovation,

Harvard Kennedy School (https://ash.harvard.edu/files/ash/files/deploying-once-only-policy.pdf).

31 These could still be managed in a federated way between government agencies. The important point here, though, is that the data is only stored once.

32 Dave Guarino, X.com post, 7 December 2021: https://x.com/allafarce/status/1468240859560878081.

33 Policy in Practice. 2022. How to increase take up of unclaimed pension credit worth £1.7 billion using data. Blog post, 20 June (https://policyinpractice.co.uk/how-to-increase-take-up-of-unclaimed-pension-credit-worth-1-7-billion-using-data/).

Chapter 2. Complexity simplified

1 Melvin E. Conway. 1968. How do committees invent? *Datamation* 14(4): 28–31 (www.melconway.com/Home/pdf/committees.pdf).

2 'Behörden und Institutionen des Bundes': www.service.bund.de/Content/DE/Behoerden/Suche/Formular.html?nn=4641496&resultsPerPage=100&cl2Categories_Einordnung=oberebundesbehoerde+unterebundesbehoerde+mittlerebundesbehoerde+oberstebundesbehoerde (accessed 27 June 2024). 'The German Federal Government': www.deutschland.de/en/topic/politics/the-german-federal-government. 'Departments, agencies and public bodies': www.gov.uk/government/organisations. 'Public entities of South Africa': https://nationalgovernment.co.za/units/type/6/public-entity.

3 Juozas Kaziukėnas. 2024. Amazon added 5 million sellers since 2018. *Marketplace Pulse*, 23 April (www.marketplacepulse.com/articles/amazon-added-5-million-sellers-since-2018).

4 David Harvey. 1990. Between space and time: reflections on the geographical imagination. *Annals of the Association of American Geographers* 80(3): 418–34 (https://doi.org/10.1111/j.1467-8306.1990.tb00305.x). See this for a short description: The Curious Geographer. 2020. What is time space compression? A Level geography in a minute. YouTube, 22 June (www.youtube.com/watch?v=vC_o5vJrQvI). See this for how the thinking is applied to more modern digital platforms: Robert Hassan. 2020. The condition of digitality: a new perspective on time and space. In *The Condition of Digitality: A Post-Modern Marxism for the Practice of Digital Life*, chapter 4. University of Westminster Press (www.jstor.org/stable/j.ctvw1d5k0.6).

5 GeoDZ, 'Time–space distanciation': www.geodz.com/eng/d/time-space-distanciation/time-space-distanciation.htm.

6 Government Digital Service. 2013. 24 departments later. Government Digital Service blog, 30 April (https://gds.blog.gov.uk/2013/04/30/24-departments-later/).

7 Nick Tait, Alan Wright and Jennifer Allum. 2018. GOV.UK: a journey in scaling agile. Government Digital Service blog, 26 April (https://gds.blog.gov.uk/2018/04/26/gov-uk-a-journey-in-scaling-agile/).

8 Sam Trendall. 2019. 'We have only scratched the surface' – Estonia's CIO on what's next for the world's most celebrated digital nation. *PublicTechnology*, 18 February (www.publictechnology.net/2019/02/18/education-and-skills/we-have-only-scratched-surface-estonias-cio-whats-next-worlds-most/).

9 Government Technology Agency of Singapore. 2019. The tech behind the Moments of Life (Families) app. Press release, 23 January (www.tech.gov.sg/media/technews/the-tech-behind-the-moments-of-life).

10 *Infra*, 'The Italian way to mobility as a service': www.infrajournal.com/en/w/the-italian-way-to-mobility-as-a-service-2.

11 Maurice V. Wilkes, David J. Wheeler and Stanley Gill. 1951. *The Preparation of Programs for an Electronic Digital Computer*. Addison-Wesley (https://archive.org/details/programsforelect00wilk/mode/2up). Carnegie Mellon University Libraries, 'New to Special Collections: the first published book on computer programming': www.library.cmu.edu/about/news/2022-12/first-computer-programming-book. Joshua Bloch. 2018. A brief, opinionated history of the API. Presentation delivered at QCon, 8 August (www.infoq.com/presentations/history-api/). Association for Computer Machinery, 'Maurice V. Wilkes: A. M. Turing Award laureate': https://amturing.acm.org/award_winners/wilkes_1001395.cfm.

12 API Evangelist, 'History of APIs': https://apievangelist.com/info/history/.

13 'FlickrUp for Nokia', Flickr app: www.flickr.com/services/apps/72157632179363369/. 'Windows Photo Gallery & Movie Maker', Flickr app: www.flickr.com/services/apps/72157601769320020/. Cal Henderson. 2008. Ten reasons to love Web 2.0. SlideShare, 7 September (www.slideshare.net/iamcal/ten-reasons-to-love-web-20-presentation).

14 'Screens around town: Moo.com Flickr MiniCards': https://signalvnoise.com/archives2/screens_around_town_moocom_flickr_minicards. 'Bubblr', Flickr app: www.flickr.com/services/apps/72157622507142125/.

15 Caterina Fake. 2006. BizDev 2.0. Blog post, 16 August (https://web.archive.org/web/20101201222906/http://caterina.net/archive/000996.html).

16 Aneesh Chopra and Nick Sinai. 2015. Wholesale
    government: open data and APIs. *Shorenstein Center,*
    Medium, 9 April (https://medium.com/@ShorensteinCtr/
    wholesale-government-open-data-and-apis-7d5502f9e2b).

17 Lindsey J. Smith. 2019. Booking campsites on Recreation.gov is a
    mess. Here's the solution. *San Francisco Chronicle,* 4 March (www.
    sfchronicle.com/travel/article/Recreation-gov-is-a-mess-Here-s-the-
    solution-13655884.php).

18 Just Eat, 'Food safety and hygiene ratings': www.just-eat.co.uk/help/
    article/207150589/food-safety-and-hygiene-ratings.

19 Food Standards Agency, 'Search for food hygiene ratings': https://
    ratings.food.gov.uk/.

20 HM Revenue & Customs, 'Find software for making tax digital for
    VAT': www.tax.service.gov.uk/making-tax-digital-software.

21 DigitalVA. 2019. Benefits intake API reaches new weekly milestone of
    900 form submissions. Press release, 21 June (https://digital.va.gov/
    general/benefits-intake-api-reaches-new-weekly-milestone-of-900-
    form-submissions/).

22 Postel + Internet Engineering Task Force. 1981. Transmission control
    protocol: DARPA internet program protocol specification. University
    of Southern California, September (www.ietf.org/rfc/rfc793.txt).

23 Olivia Neal (host). 2023. Ukraine's digital identity approach.
    Episode 46 of the *Public Sector Future* podcast (https://wwps.
    microsoft.com/episodes/ukraine-digital-government). Justin Petrone.
    2021. Deployment of Trembita system in Ukraine a milestone for
    Estonian digitisation efforts. E-Estonia, 22 April (https://e-estonia.
    com/deployment-of-trembita-system-in-ukraine-a-milestone-for-
    estonian-digitisation-efforts/).

24 Jessica Murray and Robyn Vinter. 2024. 'We milked the hell out of
    it': what happens after local food places go viral? *The Guardian,*
    23 February (www.theguardian.com/technology/2024/feb/23/
    we-milked-the-hell-out-of-it-what-happens-after-local-food-places-go-
    viral).

25 John Simmonds, Judith Harwin, Rebecca Brown and Karen
    Broadhurst. 2019. Special guardianship: a review of the evidence;
    summary report. Nuffield Family Justice Observatory (www.
    nuffieldfjo.org.uk/wp-content/uploads/2021/05/NuffieldFJO-Special-
    Guardianship-190731-WEB-final.pdf).

26 Mike Bracken. 2019. Argentina just made driving licences digital.
    Public Digital, 13 February (https://public.digital/2019/02/12/
    argentina-just-made-driving-licences-digital). Richard Pope.
    2019. Playbook: government as a platform. Ash Center

for Democratic Governance and Innovation, Harvard Kennedy School (https://ash.harvard.edu/wp-content/uploads/2024/02/293091_hvd_ash_gvmnt_as_platform_v2.pdf).

27 Amazon, 'Leadership principles': www.amazon.jobs/content/en-gb/our-workplace/leadership-principles.

28 White House. 2021. Executive order on transforming federal customer experience and service delivery to rebuild trust in government. Press release, 13 December (www.whitehouse.gov/briefing-room/presidential-actions/2021/12/13/executive-order-on-transforming-federal-customer-experience-and-service-delivery-to-rebuild-trust-in-government/). 'Life Experience Designation Charter: recovering from a disaster': https://assets.performance.gov/cx/files/life-experiences/2022/CX-2022-Life-Experience-Charter_Disaster-Recovery.pdf.

Chapter 3. Certainty over time

1   David M. Sanbonmatsu, David L. Strayer, Nathan Medeiros-Ward and Jason M. Watson. 2013. Who multi-tasks and why? Multi-tasking ability, perceived multi-tasking ability, impulsivity, and sensation seeking. *PLOS ONE* 8(1): paper e54402 (https://doi.org/10.1371/journal.pone.0054402). Melina R. Uncapher, Monica K. Thieu and Anthony D. Wagner. 2016. Media multitasking and memory: differences in working memory and long-term memory. *Psychonomic Bulletin and Review* 23(2): 483–90 (https://doi.org/10.3758/s13423-015-0907-3).

2   Earl Miller. 2017. Multitasking: why your brain can't do it and what you should do about it. Slides from a presentation delivered at MIT, 11 April (https://radius.mit.edu/sites/default/files/images/Miller%20Multitasking%202017.pdf).

3   Steve Krug. 2000. *Don't Make Me Think: A Common Sense Approach to Web Usability.* New Riders.

4   Tim Paul. 2015. One thing per page. Design in Government blog, 3 July (https://designnotes.blog.gov.uk/2015/07/03/one-thing-per-page/).

5   David A. Mindell. 2008. *Digital Apollo: Human and Machine in Spaceflight.* MIT Press.

6   Lori Cameron. 2016. Margaret Hamilton: first software engineer. IEEE Computer Society, 10 May (www.computer.org/publications/tech-news/events/what-to-know-about-the-scientist-who-invented-the-term-software-engineering). David L. Chandler. 2019. Behind the scenes of the Apollo Mission at MIT. *MIT News*, 18 July (https://news.mit.edu/2019/behind-scenes-apollo-mission-0718).

7   David Brock. 2017. Oral history of Margaret Hamilton. Computer History Museum (https://archive.computerhistory.org/resources/access/text/2022/03/102738243-05-01-acc.pdf).

8 Brian Hayes. 2019. Moonshot computing. *American Scientist*, May–June (www.americanscientist.org/article/moonshot-computing). Margaret Hamilton. 2019. The Apollo on-board flight software. March (https://wehackthemoon.com/sites/default/files/2019-03/mhh. software.final-2_0.pdf).

9 Terry Farrell and Adam Nathaniel. 2019. *Revisiting Postmodernism*. RIBA Publishers. John Higgs. 2015. *Stranger Than We Can Imagine: Making Sense of the Twentieth Century*. Signal.

Chapter 4. Shards of identity

1 John Simpson. 2009. *A Mad World, My Masters: Tales from a Traveller's Life*. Pan Macmillan.

2 'Request your personal information from the Department for Work and Pensions': www.gov.uk/guidance/request-your-personal-information-from-the-department-for-work-and-pensions#if-you-need-proof-of-a-benefit-claim.

3 Ministry of Digital Transformation (Ukraine). 2021. Ministry of Digital Transformation: Ukraine is the first country in the world to fully legalize digital passports in smartphones. Press release, 30 March (www.kmu.gov.ua/en/news/mihajlo-fedorov-ukrayina-persha-derzhava-svitu-v-yakij-cifrovi-pasporti-u-smartfoni-stali-povnimi-yuridichnimi-analogami-zvichajnih-dokumentiv). Diia. 2022. Digital passport in Diia. YouTube, 24 May (www.youtube.com/watch?v=LMmk0HSZgDI). E-Governance Academy. 2021. Diia mobile application evaluation report (https://eu4digitalua.eu/wp-content/uploads/2021/12/diia-evaluation-report.pdf).

4 Tom Phillips. 2023. Eight in ten drivers in New South Wales now store their driving licence on their smartphone. *NFCW*, 6 February (www.nfcw.com/2023/02/06/381814/eight-in-ten-drivers-in-new-south-wales-now-store-their-driving-licence-on-their-smartphone/). Mongkol Bangprapa. 2023. Digital ID now accepted to board domestic flights. *Bangkok Post*, 3 February (www.bangkokpost.com/thailand/general/2498366/digital-id-now-accepted-to-board-domestic-flights). Randy Barrett. 2023. Digital driver's licenses are finding their way to state and federal agencies. *FedTech*, 8 February (https://fedtechmagazine.com/article/2023/02/digital-drivers-licenses-are-finding-their-way-state-and-federal-agencies). Zane Lilley. 2023. France to trial digital driving licences as move online gathers pace. *The Connexion*, 13 June (www.connexionfrance.com/news/france-to-trial-digital-driving-licences-as-move-online-gathers-pace/144910). Mike Bracken. 2019. Argentina just made driving licences digital. Public Digital, 13 February (https://public.digital/2019/02/12/argentina-just-made-driving-licences-digital). *Telecompaper*. 2019. Norwegian

govt plans optional digital driving licence. *Telecompaper*, 7 March (www.telecompaper.com/news/norwegian-govt-plans-optional-digital-driving-licence--1283736). Angelica Mari. 2019. Demand for digital driving licenses soars in Brazil. *ZDNET*, 4 March (www.zdnet.com/article/demand-for-digital-driving-licenses-soars-in-brazil/).

5   Nancy Norris and Aaron Unger. 2023. Digital trust: for the people and for the planet. *The Exchange*, Medium, 5 July (https://medium.com/exchange-bc/digital-trust-for-the-people-and-for-the-planet-66fc3ef79d0b).

6   Suprita Anupam. 2023. Union budget 2023: entity DigiLocker for MSMEs, startups. *Inc42 Media*, 1 February (https://inc42.com/buzz/union-budget-2023-entity-digilocker-for-msmes-startups/).

7   Joe Rossignol. 2022. These 9 US states will let you add your driver's license to your iPhone. *MacRumors*, 9 November (www.macrumors.com/2022/10/07/iphone-wallet-ids-which-states/). Google Wallet, 'Store your digital ID on your phone': https://wallet.google/digitalid/.

8   Joe Rossignol. 2023. Apple announces businesses can accept iPhone IDs later this year. *MacRumors*, 7 June (www.macrumors.com/2023/06/07/apple-wallet-iphone-ids-at-businesses/).

9   'EU Digital Identity Wallet home': https://ec.europa.eu/digital-building-blocks/sites/display/EUDIGITALIDENTITYWALLET/EU+Digital+Identity+Wallet+Home.

10   'Digital Locker Technology Framework: version 1.1': www.dla.gov.in/sites/default/files/pdf/DigitalLockerTechnologyFramework%20v1.1.pdf.

11   DGov Korea. 2022. National Digital Document Wallet in Korea. YouTube, 4 March (www.youtube.com/watch?v=_S87FoBdvcc).

12   Google Help. 2023. How to present your state ID or driver's license in Google Wallet at airport security. YouTube, 1 June (www.youtube.com/watch?v=aIrXNm2XKH4). Rossignol. 2023. Apple announces businesses can accept iPhone IDs later this year. Screenshot of data attributes being accessed from a Google wallet: https://developers.google.com/static/wallet/identity/images/online.svg.

13   Mike Butcher. 2007. Jaiku bought by Google. *TechCrunch*, 9 October (https://techcrunch.com/2007/10/09/jaiku-bought-by-google/).

14   Jyri Engeström. 2005. Why some social network services work and others don't – or: the case for object-centered sociality. *Zengestrom*, 13 April (www.zengestrom.com/blog/2005/04/why-some-social-network-services-work-and-others-dont-or-the-case-for-object-centered-sociality.html).

15   Karin Knorr Cetina. 2003. *Epistemic Cultures: How the Sciences Make Knowledge*. Harvard University Press. Karin Knorr Cetina, Theodore

R. Schatzki and Eike von Savigny (eds). 2001. *The Practice Turn in Contemporary Theory*. Routledge.

16 Engeström. Why some social network services work and others don't – or: the case for object-centered sociality. See also Patrick Crampton. 2007. Social objects: everything you ever wanted to know! *Gapingvoid*, 31 December (www.gapingvoid.com/social-objects-for-beginners/).

17 Jack Clayton Thompson. 2018. The rights network: 100 years of the Hohfeldian rights analytic. *Laws* 7(3): paper 28 (https://doi.org/10.3390/laws7030028).

18 Legiscan, 'LA HB481': https://legiscan.com/LA/text/HB481/id/1420649.

19 Lisa M. Austin. 2012. Privacy, shame and the anxieties of identity. SSRN (https://doi.org/10.2139/ssrn.2061748).

20 Diia. 2023. Reels at World Economic Forum 2023. YouTube, 18 January (www.youtube.com/live/Zv4HVp6UP8Y?feature=share&t=1727). France Num, 'Dématérialisation des documents': www.francenum.gouv.fr/guides-et-conseils/pilotage-de-lentreprise/dematerialisation-des-documents.

21 National E-Governance Division, 'Digi Locker': https://negd.gov.in/digilocker/.

22 'DigiLocker: an initiative towards paperless governance': www.digilocker.gov.in/about/about-digilocker.

23 Adam Vidler. 2023. Centuries-old rules for these legal documents set to change. *9News*, 7 September (www.9news.com.au/national/statutory-declarations-going-digital-new-government-legislation/60c59327-8945-46be-944b-0ffaa32b5407).

24 Joe Tomlinson, Jack Maxwell and Alice Welsh. 2021. Discrimination in digital immigration status. *Legal Studies* 42(2): 315–34 (https://doi.org/10.1017/lst.2021.33).

25 Government of British Columbia, 'Digital trust technical overview': https://digital.gov.bc.ca/digital-trust/technical/.

26 Agus Brizuela. 2023. Using a Wardley map to understand how a digital driver's license can boost digital government evolution. Medium, 9 February (https://medium.com/@agustina.b/using-a-wardley-map-to-understand-how-a-digital-drivers-license-can-boost-digital-government-5d97d82b301b). Richard Pope. 2019. Data infrastructure, APIs and open standards. In *Playbook: Government as a Platform*. Ash Center for Democratic Governance and Innovation, Harvard Kennedy School (https://ash.harvard.edu/wp-content/uploads/2024/02/293091_hvd_ash_gvmnt_as_platform_v2.pdf).

Chapter 5. Common components

1   Datalink Article – 'A Licence to Process' – Computing at the DVLC, c. 1978. Centre for Computing History (www.computinghistory.org. uk/det/42542/Datalink-Article-A-Licence-to-Process-Computing-at-the-DVLC-c-1978/).

2   Ward Cunningham. 2009. Debt metaphor. YouTube, 15 February (www.youtube.com/watch?v=pqeJFYwnkjE).

3   Will Myddelton and Ciara Green. 2017. Discovering the next government as a platform component. Government as a Platform blog, 6 September (https://governmentasaplatform.blog.gov. uk/2017/09/06/discovering-the-next-component/).

4   TheyWorkForYou, 'Driving licences: applications': www.theyworkforyou.com/wrans/?id=2021-09-07. HL2506.h&s=DVLA+lorries+mail#gHL2506.r0. TheyWorkForYou, 'DVLA response time': www.theyworkforyou.com/ wrans/?id=2000-01-11a.109.3&s=DVLA+lorries+mail#g109.5.

5   US Government Accountability Office. 2021. Tax filing: actions needed to address processing delays and risks to the 2021 filing season. Report, 1 March (www.gao.gov/products/gao-21-251). Lori Hawkins. 2022. IRS reverses course, says it won't close Austin processing facility. Austin American-Statesman, 18 February (https:// eu.statesman.com/story/business/2022/02/18/irs-backlog-tax-returns-reverses-plan-close-austin-tx-office/6836211001/). Matthew Busch. 2022. Opinion: why does the IRS need $80 billion? Just look at its cafeteria. Washington Post, 9 August (www.washingtonpost.com/ opinions/interactive/2022/irs-pipeline-tax-return-delays/).

6   Miles Brignall and Rebecca Smithers. 2020. Motorists in a jam as Covid-19 leaves them waiting months for DVLA documents. The Guardian, 1 August (www.theguardian.com/money/2020/aug/01/ motorists-covid-19-dvla-documents-car-application).

7   Tim O'Reilly. 2011. Government as a platform. Innovations: Technology, Governance, Globalization 6(1): 13–40 (https://doi. org/10.1162/inova000 56).

8   Eric Steven Raymond. 2003. Basics of the Unix philosophy. In The Art of Unix Programming, chapter 1, section 6. Published online (www. catb.org/esr/writings/taoup/html/ch01s06.html).

9   Tim Paul. 2016. Introducing GOV.UK Frontend alpha. Design in Government blog, 14 October (https://designnotes.blog.gov. uk/2016/10/14/introducing-gov-uk-frontend-alpha/).

10 Till Wirth. 2015. Introducing GOV.UK Pay. Government Digital Service blog, 15 October (https://gds.blog.gov.uk/2015/10/15/introducing-gov-uk-pay/). Government Digital Service, 'A GDS story 2016': https://

gds.blog.gov.uk/story-2016/. Pete Herlihy, X.com post, 16 April 2019: https://twitter.com/yahoo_pete/status/1118245112482598912.

11   Josh Lowe. 2018. What a Big Tech executive learned in government. *Apolitical*, 12 November (https://apolitical.co/solution-articles/en/what-a-big-tech-executive-learned-transforming-italys-government).

12   Government of British Columbia, 'Digital trust': https://digital.gov.bc.ca/digital-trust/home/. Government of British Columbia, 'Common components': https://digital.gov.bc.ca/common-components/.

13   Tim Paul. 2020. Measuring the value of the GOV.UK Design System. Talk given at the 14th Gov Design Meetup, 5 February (www.youtube.com/watch?v=eSkVtSEAe98).

14   Cabinet Office, Government Digital Service and the Rt Hon. Simon Hart. 2019. Government's streamlined messaging service to save taxpayer £175m. Press release, 26 September (www.gov.uk/government/news/governments-streamlined-messaging-service-to-save-taxpayer-175m).

15   Anais Reding, Chris Pinder and Kaz Hufton. 2018. How government as a platform is making things better for users. Government as a Platform blog, 13 March (https://governmentasaplatform.blog.gov.uk/2018/03/13/making-things-better-for-users/).

16   Richard Pope. 2019. Government as a platform, the hard problems: part 2 – the design of public-facing services. Medium, 6 August (https://medium.com/@richardjpope/government-as-a-platform-the-hard-problems-part-2-the-design-of-public-facing-services-4b3b2447f379).

17   'Centralization gone right: a case study on the US Web Design System': https://github.com/18F/HCD_for_IT_Centralization/blob/master/case_study_USWDS.md.

18   Mike Bracken. 2019. Argentina just made driving licences digital. Public Digital, 13 February (https://public.digital/2019/02/12/argentina-just-made-driving-licences-digital).

19   Jen Allum. 2020. Update on the future of GOV.UK. Government Digital Service blog, 28 May (https://gds.blog.gov.uk/2020/05/28/update-on-the-future-of-gov-uk/). Gavin Freeguard, Marcus Shepheard and Oliver Davies. 2020. Digital government during the coronavirus crisis. Institute for Government, 18 November (www.instituteforgovernment.org.uk/publication/report/digital-government-during-coronavirus-crisis). Mark Reynolds. 2020. How we created the shielded patient list. NHS Digital, 17 November (https://digital.nhs.uk/blog/tech-talk/2020/how-we-created-the-shielded-patient-list). GDS Team. 2021. Podcast: the Clinically Extremely Vulnerable People service. Government Digital

Service blog, 24 February (https://gds.blog.gov.uk/2021/02/24/podcast-the-clinically-extremely-vulnerable-people-service/).

20 Marcel Saulnier. 2020. Get updates on Covid-19: email notification service. Canadian Digital Service, 14 May (https://digital.canada.ca/2020/05/13/get-updates-on-covid-19--email-notification-service/).

21 Makena Kelly. 2020. Unemployment checks are being held up by a coding language almost nobody knows. *The Verge*, 14 April (www.theverge.com/2020/4/14/21219561/coronavirus-pandemic-unemployment-systems-cobol-legacy-software-infrastructure).

22 Michele Evermore. 2023. New Jersey's worker-centered approach to improving the administration of unemployment insurance. Heldrich Center for Workforce Development, Rutgers University (https://heldrich.rutgers.edu/sites/default/files/2023-09/New_Jersey%E2%80%99s_Worker-centered_Approach_to_Improving_the_Administration_of_Unemployment_Insurance.pdf). Daniel Munoz. 2023. NJ unemployment claim issues? They may be fixed with these new funds. *North Jersey Media Group*, September 28 (https://eu.northjersey.com/story/money/workplace/2023/09/28/nj-unemployment-claims-status-federal-funds-easier/70993194007/).

23 Andrew Mackley. 2021. Coronavirus: Universal Credit during the crisis. House of Commons Library, 15 January (https://researchbriefings.files.parliament.uk/documents/CBP-8999/CBP-8999.pdf). Neither identity system was ideal, but that it could be changed is what is significant here.

24 Derek du Preez. 2020. How DWP managed a surge in demand for Universal Credit during Covid-19. *Diginomica*, 13 November (https://diginomica.com/how-dwp-managed-surge-demand-universal-credit-during-covid-19).

25 Committee of Public Accounts. 2022. Department for Business, Energy and Industrial Strategy annual report and accounts 2020–21. Report, 18 May (https://publications.parliament.uk/pa/cm5803/cmselect/cmpubacc/59/report.html#heading-1).

26 Richard Pope. 2020. The UK government should negotiate free access to faster payments to speed up Covid-19 payments. Blog post, 15 April (https://richardpope.org/blog/2020/04/15/faster-payments-covid-19/). BBC. 2020. Coronavirus: Trump's name to appear on US relief cheques. *BBC News*, 15 April (www.bbc.co.uk/news/world-52293910).

27 *The Economist*. 2020. Covid-19 spurs national plans to give citizens digital identities. *The Economist*, 7 December (www.economist.com/international/2020/12/07/covid-19-spurs-national-plans-to-give-citizens-digital-identities).

28 Marie Hicks. 2018. *Programmed Inequality*. MIT Press.

29 Simon Sharwood. 2023. G20 digital ministers sign up for digital public infrastructure push. *The Register*, 21 August (www.theregister. com/2023/08/21/g20*digital*publicin*frastructure*plan/).

30 MOSIP. 2023. MOSIP and Sierra Leone sign MoU to pilot national digital ID system. Press release, 10 January (https://mosip.io/ news_events/mosip-and-sierra-leone-sign-mou-to-pilot-national-digital-id-system). MOSIP. 2022. MOSIP enters partnership with Burkina Faso. Press release, 20 December (https://mosip.io/ news_events/mosip-enters-partnership-with-burkina-faso). MOSIP. 2022. The government of Madagascar signs an MoU for a MOSIP pilot programme. Press release, 4 November (https://mosip.io/ news_events/the-government-of-madagascar-signs-an-mou-for-a-mosip-pilot-programme).

31 Sharwood. 2023. G20 digital ministers sign up for digital public infrastructure push. GitHub Social Impact, 'DPG Open Source Community Manager Program': https://socialimpact.github.com/ tech-for-social-good/dpg-open-source-community-manager-program.

32 Pete Herlihy. 2020. UK GDS: GOV.UK Notify. *Project on Digital Era Government*, Medium, 8 June (https://medium.com/ digitalhks/u-k-gds-gov-uk-notify-e645cce3cda8).

33 Richard Pope. 2019. Interview with Will Myddelton: UK Government as a Platform programme. *Project on Digital Era Government*, Medium, 29 July (https://medium.com/platform-land/interview-with-will-myddelton-government-as-a-platform-3aff4ebcb3e8).

34 Richard Pope. 2019. Identifying platforms. In *Playbook: Government as a Platform*. Ash Center for Democratic Governance and Innovation, Harvard Kennedy School (https://ash.harvard.edu/wp-content/ uploads/2024/02/293091_hvd_ash_gvmnt_as_platform_v2.pdf).

35 MyUSA GitHub repository: https://github.com/18F/myusa.

36 Nandan Nilekani and Viral Shah. 2015. *Rebooting India*. Allen Lane.

37 The Exchange. 2022. DigitalBC livestream fixing digital funding. YouTube, 2 November (www.youtube.com/ watch?v=w4qMSoIr6hk&list=PL9CV).

38 Richard Pope and Antonio Weis. 2021. Government as a platform: who pays? Bennett Institute for Public Policy, 15 February (www. bennettinstitute.cam.ac.uk/blog/government-platform-who-pays/).

Chapter 6. Data as infrastructure

1 GoCognitive, 'Alan Baddeley: the working memory model': www.gocognitive.net/episode/postal-codes. 'Using PAF data yourself': www.poweredbypaf.com/using-our-address-data/ use-the-data-yourself/.

2   Public Administration Committee. 2014. Open data and economic growth. In 'Statistics and open data: harvesting unused knowledge, empowering citizens and improving public services', chapter 3 (https://publications.parliament.uk/pa/cm201314/cmselect/cmpubadm/564/56406.htm).

3   www.legislation.gov.uk/all?text=%22a%20register%20of%22

4   'General Law – part I, title XXII, chapter 156D, section 2.01': https://malegislature.gov/Laws/GeneralLaws/PartI/TitleXXII/Chapter156D/Section2.01.

5   'Housing (Wales) Act 2014': www.legislation.gov.uk/anaw/2014/7/section/14/enacted.

6   HM Land Registry, 'About us': www.gov.uk/government/organisations/land-registry/about.

7   'Land Registration Act 2002': www.legislation.gov.uk/ukpga/2002/9/contents/enacted.

8   HM Revenue & Customs. 2019. Guidance on RTI data items from April 2019 (https://assets.publishing.service.gov.uk/government/uploads/system/uploads/attachment_data/file/775920/Data_items_guide_2019_to_2020_v1.1.pdf). 'Real time information': http://data.parliament.uk/DepositedPapers/Files/DEP2019-0465/Real_Time_Information_v6.0.pdf. 'HMRC and data sharing with the DWP': www.whatdotheyknow.com/request/hmrc_and_data_sharing_with_the_d. Child Poverty Action Group. 2019. Universal credit claimants blocked from challenging DWP decisions. Blog post, 19 July (https://web.archive.org/web/20220319035602/https://cpag.org.uk/news-blogs/news-listings/universal-credit-claimants-blocked-challenging-dwp-decisions).

9   'Why should I care what color the bikeshed is?': https://bikeshed.org/. C. Northcote Parkinson. 1957. *Parkinson's Law and Other Studies in Administration*. Houghton Mifflin. 'The bikeshed email': http://phk.freebsd.dk/sagas/bikeshed/.

10 E. F. Codd. 1970. A relational model of data for large shared data banks. *Communications of the ACM* 13(6): 377–87 (https://doi.org/10.1145/362384.362685).

11 IBM, 'The relational database': www.ibm.com/history/relational-database. *Two-Bit History*. 2017. Important papers: Codd and the relational model. Blog post, 29 December (https://twobithistory.org/2017/12/29/codd-relational-model.html).

12 National Research Council. 1999. *Funding a Revolution: Government Support for Computing Research*. National Academies Press (https://nap.nationalacademies.org/read/6323/chapter/8#166).

13 Association for Computing Machinery, 'Edgar F. Codd: A. M. Turing Award laureate': https://amturing.acm.org/award_winners/codd_1000892.cfm.

14 Riigi Teataja, 'Public Information Act': www.riigiteataja.ee/en/
    eli/514112013001/consolide.

15 Anett Numa. 2020. Discovering the most widely used
    e-services. E-Estonia, 8 January (https://e-estonia.com/
    speakers-corner-most-widely-used-e-services/).

16 Estonian State Portal, 'Income tax and the declaration of income':
    www.eesti.ee/en/money-and-property/taxation-and-currency/
    income-tax-and-the-declaration-of-income.

17 Peeter Vihma. 2023. Beyond life and livestock: exploring Estonia's
    quirky digital services. E-Estonia, 28 July (https://e-estonia.com/
    beyond-life-and-livestock-exploring-estonias-quirky-digital-
    services/).

18 Ministry of Rural Development (India). 2023. Bhumi Samvaad IV:
    National Conference on Digitizing and Georeferencing India with
    Bhu-Aadhaar (ULPIN). Press release, 10 March (https://pib.gov.in/
    PressReleaseIframePage.aspx?PRID=1905655). Ministry of Rural
    Development (India). 2023. Year end review 2023: achievement of
    the Department of Land Resources (Ministry of Rural Development).
    Press release, 22 December (https://pib.gov.in/PressReleasePage.
    aspx?PRID=1989671). Moushumi Das Gupta. 2023. What is Bhu-
    Aadhaar? A unique ID for land that aims to check fraud, benami
    transactions. *ThePrint*, 28 March (https://theprint.in/india/
    governance/what-is-bhu-aadhaar-a-unique-id-for-land-that-aims-to-
    check-fraud-benami-transactions/1475189/).

19 National Health Authority, 'Ayushman Bharat Digital Mission': https://
    abdm.gov.in/.

20 National Health Authority, 'Ayushman Bharat Digital Mission: Health
    Facility Registry': https://facility.ndhm.gov.in.

21 Sourabh Lele. 2023. Govt, AWS working on health IDs, will maintain
    digital health records. *Business Standard*, 11 October (www.
    business-standard.com/india-news/govt-aws-working-on-health-ids-
    will-maintain-digital-health-records-123101101099_1.html). Deepto
    Banerjee. 2023. APAAR: all you need to know about this 'one nation,
    one ID' to be rolled out for students. *Times of India*, 16 October
    (https://timesofindia.indiatimes.com/education/news/apaar-all-you-
    need-to-know-about-this-one-nation-one-id-to-be-rolled-out-for-
    students/articleshow/104468708.cms).

22 Ministry of Communications (India). 2021. Digital Address Code:
    Approach Paper. Consultative draft, 18 October (https://prsindia.org/
    files/parliamentry-announcement/2021-11-20/IP_Apd_18oct21_c.pdf).

23 National Portal of India, 'Ayushman Bharat Digital Mission': www.
    india.gov.in/spotlight/ayushman-bharat-digital-mission-abdm.

24 Indian Institute of Science. 2023. Agricultural Data Exchange (ADeX)
    launched in Hyderabad. Press release, 11 August (https://iisc.ac.in/

events/agricultural-data-exchange-adex-launched-in-hyderabad/).
Indian Urban Data Exchange, 'Platform': https://iudx.org.in/platform/.

25 Harle Pihlak. 2019. First EU citizens using ePrescriptions in
other EU country. E-Estonia, 21 January (https://e-estonia.com/
first-eu-citizens-using-eprescriptions-in-other-eu-country/).

26 Ministry of Economic Affairs and Communication (Estonia), 'Real-time
economy': https://realtimeeconomy-bsr.eu/.

27 E-Estonia, 'X-Road': https://e-estonia.com/
solutions/x-road-interoperability-services/x-road/.

28 Open Network For Digital Commerce. 2024. ONDC records over
7.5 million transactions in February, up 15% from previous month.
Blog post, 4 March (https://ondc.org/blog/ondc-records-over-7-5-
million-transactions-in-february-up-15-from-previous-month/). Open
Network For Digital Commerce, 'Network Participants on ONDC':
https://ondc.org/network-participants/. Ministry of Commerce
and Industry (India). 2022. ONDC network starts beta testing with
consumers in Bengaluru. Press release, 30 September (https://
pib.gov.in/PressReleasePage.aspx?PRID=1863871). *ETGovernment*.
2023. Now buy subsidised tomatoes at Rs 70 per kg through Open
Network for Digital Commerce. *ETGovernment*, 22 July (https://
government.economictimes.indiatimes.com/news/digital-india/now-
buy-subsidised-tomatoes-at-rs-70-per-kg-through-open-network-for-
digital-commerce/102030048).

29 Kristjan Vassil. 2015. Estonian e-government ecosystem: foundation,
applications, outcomes. World Bank, June (https://thedocs.
worldbank.org/en/doc/165711456838073531-0050022016/original/
WDR16BPEstonianeGovecosystemVassil.pdf).

30 DLUHC Digital Planning. 2022. Introducing the planning data
platform. MHCLG Digital blog, 28 September (https://dluhcdigital.
blog.gov.uk/2022/09/28/introducing-the-planning-data-platform/).
Institute for Government. 2022. Data Bites #34: getting things
done with data in government. Event video and transcript,
5 October (www.instituteforgovernment.org.uk/event/
data-bites-34-getting-things-done-data-government).

31 'Planning Data': www.planning.data.gov.uk.

32 John Sheridan. GovCamp c. 2008.

33 'Interpretation Act 1978': www.legislation.gov.uk/ukpga/1978/30/
enacted.

Chapter 7. Empathy augmented

1   Sherry Turkle. 2015. Talk on *Reclaiming Conversation*, Google,
13 October (www.youtube.com/watch?v=awFQtX7tPoI).

2   Sherry Turkle. 2015. *Reclaiming Conversation: The Power of Talk in a Digital Age*. Penguin.

3   Jim Al Khalili (host). 2020. Peter Fonagy on a revolution in mental health care. Episode of *The Life Scientific*, BBC Radio 4, 28 January (www.bbc.co.uk/programmes/m000dpj2). Anthony Bateman and Peter Fonagy. 2006. *Mentalization-Based Treatment for Borderline Personality Disorder: A Practical Guide*. Oxford University Press (https://oxfordmedicine.com/view/10.1093/med/9780198570905.001.0001/med-9780198570905).

4   Blake Lee-Whiting. 2023. A rise in self-service technologies may cause a decline in our sense of community. *The Conversation*, 25 April (https://theconversation.com/a-rise-in-self-service-technologies-may-cause-a-decline-in-our-sense-of-community-201339).

5   Tom R. Tyler. 2003. Procedural justice, legitimacy, and the effective rule of law. *Crime and Justice* 30(1): 283–357 (https://doi.org/10.1086/652233).

6   Donald Moynihan, Pamela Herd and Hope Harvey. 2014. Administrative burden: learning, psychological, and compliance costs in citizen-state interactions. *Journal of Public Administration Research and Theory* 25(1): 43–69 (https://doi.org/10.1093/jopart/muu009).

7   'Administrative Fairness Lab', archived 3 October 2022: https://web.archive.org/web/20221003165717/https://www.york.ac.uk/law/research/administrative-fairness-lab/.

8   Joe Tomlinson, Jed Meers and Simon Halliday. 2024. Administrative fairness in the digital welfare state (report no. 1): procedural legitimacy logics within the digital welfare state. Administrative Fairness Lab, York Law School, 24 February (https://pure.york.ac.uk/portal/en/publications/administrative-fairness-in-the-digital-welfare-state-report-no-1-).

9   Desmond King. 1995. *Actively Seeking Work? The Politics of Unemployment and Welfare Policy in the United States and Great Britain*. Chicago University Press.

10 Michael Laffan. 2012. Universal Jobmatch: update for members; recent information from the union in DWP. PCS website, 3 December 2012 (https://web.archive.org/web/20130207090129/https://www.pcs.org.uk/en/departmentforworkandpensionsgroup/dwp-news.cfm/id/D34395B0-26B7-4E67-81B32F80CFEBB3E8).

11 'Universal Automation', archived 20 October 2013: https://web.archive.org/web/20131020164827/http://automation.strikenow.org.uk/.

12 Rowena Mason. 2014. Government's Universal Jobmatch website 'bedevilled with fraud'. *The Guardian*,

5 March (www.theguardian.com/society/2014/mar/05/ government-universal-jobmatch-website-fraud).

13 Shiv Malik. 2014. DWP draws up plans to ditch ridiculed jobs website. *The Guardian*, 16 March (www.theguardian.com/money/2014/mar/16/ dwp-jobs-website-universal-jobsmatch).

14 Doreen Massey. 1994. *Space, Place, and Gender*. University Of Minnesota Press.

15 Sophie Lovell. 2024. *Dieter Rams*. Phaidon.

16 Kevin Deldycke, 'Awesome falsehood': https://github.com/kdeldycke/ awesome-falsehood.

17 Marie Hicks. 2019. Hacking the cis-tem: transgender citizens and the early digital state. *IEEE Annals of the History of Computing* 41(1): 20–33 (https://doi.org/10.1109/MAHC.2019.2897667).

18 Cory Doctorow. 2023. Pluralistic: Netflix wants to chop down your family tree. *Pluralistic*, 2 February (https://pluralistic.net/2023/02/02/ nonbinary-families). Terence Eden. 2017. Falsehoods programmers believe about families. Blog post, 29 March (https://shkspr.mobi/ blog/2017/03/falsehoods-programmers-believe-about-families/).

19 Ozgun Atasoy and Carey K. Morewedge. 2017. Digital goods are valued less than physical goods. *Journal of Consumer Research* 44(6): 1343–57 (https://doi.org/10.1093/jcr/ucx102).

20 Rosie Mears and Sophie Howes. 2023. You reap what you code. Child Poverty Action Group, 27 June (https://cpag.org.uk/sites/default/files/ files/policypost/You_Reap_What_You_Code.pdf).

21 Wendy De La Rosa, Eesha Sharma, Stephanie M. Tully, Eric Giannella and Gwen Rino. 2021. Psychological ownership interventions increase interest in claiming government benefits. *Proceedings of the National Academy of Sciences* 118(35): paper e2106357118 (https://doi. org/10.1073/pnas.2106357118).

22 *Design Journal*. 1972. Conference visionaries. *Design Journal*, no. 274, October (www.vads.ac.uk/digital/collection/DIAD/id/5555/ rec/1). Post Office Telecommunications. N.d. Confravision (www. samhallas.co.uk/repository/journals/Descriptive_Leaflets/DLB%20 202%20Confravision%2007-75.pdf). British Movietone. 1967. Confravision. YouTube video, 2 November (www.youtube.com/ watch?v=EVC8RPJfjEk).

23 John Short, Ederyn Williams and Bruce Christie. 1976. *The Social Psychology of Telecommunications*. Wiley.

24 Karel Kreijns, Kate Xu and Joshua Weidlich. 2021. Social presence: conceptualization and measurement. *Educational Psychology Review* 34(June): 139–70 (https://doi.org/10.1007/s10648-021-09623-8). Catherine S. Oh, Jeremy N. Bailenson and Gregory F. Welch. 2018. A systematic review of social presence: definition, antecedents,

and implications. *Frontiers in Robotics and AI* 5(114) (https://doi.org/10.3389/frobt.2018.00114).

25   Amanda Ruggeri. 2024. The surprising promise and profound perils of AIs that fake empathy. *New Scientist*, 6 March (www.newscientist.com/article/mg26134810-900-the-surprising-promise-and-profound-perils-of-ais-that-fake-empathy/). Sherry Turkle. 2020. That chatbot I've loved to hate. *MIT Technology Review*, 18 August (www.technologyreview.com/2020/08/18/1006096/that-chatbot-ive-loved-to-hate/).

26   Rainer Kattel and Ville Takala. 2021. Dynamic capabilities in the public sector: the case of the UK's Government Digital Service. UCL Institute for Innovation and Public Purpose, 8 January (www.ucl.ac.uk/bartlett/public-purpose/publications/2021/jan/dynamic-capabilities-public-sector-case-uks-government-digital-service).

27   Colin Reckons. 2015. Bret Victor: media for thinking the unthinkable. YouTube, 16 June (www.youtube.com/watch?v=oUaOucZRlmE). Sherry Turkle. 2005. *The Second Self: Computers and the Human Spirit*, 20th anniversary edition. MIT Press. Sherry Turkle. 2021. *The Empathy Diaries: A Memoir*. Penguin. Sherry Turkle, 'How computers change the way we think': https://www1.udel.edu/educ/whitson/897s05/files/turkle.

28   Jenny L. Schnyder, Hanna K. de Jong, Bache E. Bache, Frieder Schaumburg and Martin P. Grobusch. 2024. Long-term immunity following yellow fever vaccination: a systematic review and meta-analysis. *The Lancet* 12(3): E445–56 (https://doi.org/10.1016/S2214-109X(23)00556-9).

Chapter 8. Designing the seams (not seamless design)

1   All about Steve Jobs.com. 2009. It just works. Seamlessly. YouTube, 19 September (www.youtube.com/watch?v=qmPq00jelpc).

2   Apple. 2023. Worldwide Developer Conference 2023 keynote (https://developer.apple.com/videos/play/wwdc2023/101/?time=5435).

3   Walter Isaacson and Steve Jobs. 2015. *Steve Jobs*. Abacus.

4   Sophie Lovell. 2024. *Dieter Rams*. Phaidon. See also 'Dasprogramm' at https://dasprogramm.co.uk.

5   Walter Isaacson. 2012. How Steve Jobs' love of simplicity fueled a design revolution. *Smithsonian Magazine*, 1 September (www.smithsonianmag.com/arts-culture/how-steve-jobs-love-of-simplicity-fueled-a-design-revolution-23868877/).

6   Sue Factor. 2008. What makes a design 'Googley'? *Official Google Blog*, 23 April (https://googleblog.blogspot.com/2008/04/what-makes-design-googley.html).

7   Amazon, 'Leadership principles': www.aboutamazon.com/about-us/
    leadership-principles.
8   David Allen Green. 2024. Why it's time to drop the old
    doctrine of ministerial responsibility. *Prospect*, 25 January
    (www.prospectmagazine.co.uk/ideas/law/law-and-
    government/64609/drop-doctrine-ministerial-responsibility-
    david-allen-green). *Reestheskin*. 2015. What Nye Bevan
    actually said. Blog post, 30 December (https://reestheskin.me/
    what-nye-bevan-actually-said/).
9   'Digital by Default Service Standard: government service design
    manual', archived 29 August 2013: https://web.archive.org/
    web/20130829040233/https://www.gov.uk/service-manual/
    digital-by-default.
10 Roger B. Parks, Paula C. Baker, Larry Kiser, Ronald Oakerson,
    Elinor Ostrom, Vincent Ostrom, Stephen L. Percy, *et al.* 1981.
    Consumers as coproducers of public services: some economic
    and institutional considerations. *Policy Studies Journal* 9(7):
    1001–11 (https://doi.org/10.1111/j.1541-0072.1981.tb01208.x).
11 Elinor Ostrom and Gordon Whitaker. 1973. Does local community
    control of police make a difference? Some preliminary findings.
    American Journal of Political Science 17(1): 48–76 (https://doi.
    org/10.2307/2110474).
12 Paul Dragos Aligica and Vlad Tarko. 2013. Co-production,
    polycentricity, and value heterogeneity: the Ostroms' public choice
    institutionalism revisited. *American Political Science Review* 107(4):
    726–41 (https://doi.org/10.1017/s0003055413000427).
13 Catherine C. Marshall. 1998. The future of annotation in a digital
    (paper) world. Paper presented at the 35th Annual GSLIS Clinic:
    Successes and Failures of Digital Libraries, University of Illinois at
    Urbana-Champaign, 24 March.
14 Aligica and Tarko. 2013. Co-production, polycentricity, and value
    heterogeneity.
15 Vernon Bogdanor. 2015. The general election, February 1974. Lecture
    delivered at Gresham College (www.gresham.ac.uk/watch-now/
    general-election-february-1974).
16 Richard Pope. 2011. A few design rules for Alpha.gov.uk. Government
    Digital Service blog, 28 April (https://gds.blog.gov.uk/2011/04/28/
    alpha-gov-uk-design-rules/). 'GDS design principles', archived 1 May
    2012: https://web.archive.org/web/20120501210618/https://www.gov.
    uk/designprinciples.
17 Jeni Tennison. 2012. Precious snowflakes. Blog post, 10 March (www.
    jenitennison.com/2012/03/10/precious-snowflakes.html).

18 Government Digital Service. 2018. Building the GOV.UK of the future. Blog post, 27 June (https://gds.blog.gov.uk/2018/06/27/building-the-gov-uk-of-the-future/).

19 Joshua Rozenberg. 2019. Justice online: are we there yet? Lecture delivered at Gresham College, 21 February (www.gresham.ac.uk/watch-now/justice-online-there-yet).

20 Kevin Lynch. 1960. *The Image of the City*. London: The MIT Press.

21 Neil Williams. 2012. A quick tour of inside government. Government Digital Service blog, 15 November (https://gds.blog.gov.uk/2012/11/15/a-quick-tour-of-inside-government/).

22 Marc Weiser. 1994. The world is not a desktop. *Interactions* 1(1): 7–8 (https://doi.org/10.1145/174800.174801).

23 Mark Weiser and John Seely Brown. 1995. Designing calm technology. Xerox PARC, 21 December (https://people.csail.mit.edu/rudolph/Teaching/weiser.pdf).

24 Sarah Inman and David Ribes. 2019. Beautiful seams: strategic revelations and concealments. In *Proceedings of the 2019 CHI Conference on Human Factors in Computing Systems*, paper 278. ACM (https://doi.org/10.1145/3290605.3300508). Otrops. 2010. Beautiful seams. Blog post, 23 April (https://otrops.com/notes/beautiful-seams/). Adam Greenfield. 2007. On the ground running: lessons from experience design. *Speedbird*, 23 June (https://speedbird.wordpress.com/2007/06/22/on-the-ground-running-lessons-from-experience-design/). 'The computer for the 21st century', archived 21 July 2013: https://web.archive.org/web/20130721194538/http://www.ubiq.com/hypertext/weiser/SciAmDraft3.

25 Matthew Chalmers and Areti Galani. 2004. Seamful interweaving. In *Proceedings of the 2004 Conference on Designing Interactive Systems Processes, Practices, Methods, and Techniques*, pp. 243–52. ACM (https://doi.org/10.1145/1013115.1013149).

26 Jeremy Keith. 2014. Seams. *Adactio*, 12 May (https://adactio.com/journal/6786).

27 Denis Campbell. 2018. NHS will no longer have to share immigrants' data with Home Office. *The Guardian*, 9 May (www.theguardian.com/society/2018/may/09/government-to-stop-forcing-nhs-to-share-patients-data-with-home-office). Alan Travis. 2018. NHS chiefs urged to stop giving patient data to immigration officials. *The Guardian*, 31 January (www.theguardian.com/society/2018/jan/31/nhs-chiefs-stop-patient-data-immigration-officials).

28 Richard Pope. 2020. Trust and accountability patterns in digital government: data usage trackers. Bennett Institute for Public Policy, 27 October (www.bennettinstitute.cam.ac.uk/blog/trust-and-accountability-patterns-digital-governme/).

29 Julinda Beqiraj, Akanksha Bisoyi, Christian Djeffal, Mark Findlay, Jane Loo and Ong Li Min. 2022. White paper: rule of law, legitimacy and effective Covid-19 control technologies. Bingham Centre for the Rule of Law, 29 July (https://binghamcentre.biicl.org/publications/white-paper-rule-of-law-legitimacy-and-effective-covid-19-control-technologies).

30 *Mystery of Existence*. 2021. Brandon Bradford: it's easy to make everything into a conspiracy when you don't know how anything works. Blog post, 19 November (https://absentofi.org/2021/11/brandon-bradford-its-easy-to-make-everything-into-a-conspiracy-when-you-dont-know-how-anything-works/).

31 'Early warning system findings': https://cpag.org.uk/policy-and-research/findings-our-projects/early-warning-system-findings.

32 Jakob Nielsen. 2006. Progressive disclosure. Nielsen Norman Group, 3 December (www.nngroup.com/articles/progressive-disclosure/).

33 Joe Tomlinson, Jed Meers and Simon Halliday. 2024. Administrative fairness in the digital welfare state (report no. 1): procedural legitimacy logics within the digital welfare state. Administrative Fairness Lab, York Law School, 24 February (https://nuffieldfoundation.org/wp-content/uploads/2022/09/Feb-2024-Publication-Version-Report-1-Administrative-Fairness-in-the-Digtial-Welfare-State.pdf).

34 Rob Kling. 1977. The organizational context of user-centered software designs. *MIS Quarterly* 1(4): 41–52 (https://doi.org/10.2307/249021).

Chapter 9. Accountable automation

1 James Tomayko. 1988. Computing in spaceflight: the NASA experience. NASA History Office (https://docs.google.com/document/d/1XTVLIkwoAEf5mZnSwRFCHaW1_PZ4gG4d15NWd2arvtA/edit). NASA Jet Propulsion Laboratory. 2023. NASA's Voyager team focuses on software patch, thrusters. Press release, 23 October (https://voyager.jpl.nasa.gov/news/details.php?article_id=131).

2 Christine Lagorio-Chafkin. 2012. Kevin Systrom and Mike Krieger, founders of Instagram. *Inc.*, 9 April (www.inc.com/30under30/2011/profile-kevin-systrom-mike-krieger-founders-instagram.html).

3 Cameron Faulkner. 2023. Apple is turning on the HomePod Mini's secret temperature and humidity sensor. *The Verge*, 18 January (www.theverge.com/2023/1/18/23560630/apple-homepod-mini-temperature-humidity-sensor-update).

4 Peter Ackroyd. 2007. *Thames: Sacred River*. Random House.

5   J. W. Franks, A. J. Sutcliffe, M. P. Kerney and G. R. Coope. 1958. Haunt of elephant and rhinoceros: the Trafalgar Square of 100,000 years ago; new discoveries. *Illustrated London News*, 14 June, pp. 1101–3. Hackney Museum Collections Online, 'Bone – woolly rhinoceros': https://museum-collection.hackney.gov.uk/object-1991-879.

6   M. M. Lehman. 1980. Programs, life cycles, and laws of software evolution. *Proceedings of the IEEE* 68(9): 1060–76 (https://doi.org/10.1109/proc.1980.11805). Gerardo Canfora, Darren Dalcher, David Raffo, Victor R. Basili, Juan Fernández-Ramil, Václav Rajlich, Keith Bennett, *et al*. 2011. In memory of Manny Lehman, 'father of software evolution'. *Journal of Software Maintenance and Evolution: Research and Practice* 23(3): 137–44 (https://doi.org/10.1002/smr.537).

7   Philip Frana. 2002. An interview with Laszlo A. Belady. Charles Babbage Institute, University of Minnesota (https://conservancy.umn.edu/bitstream/handle/11299/107110/oh352lab.pdf?sequence=1&isAllowed=y).

8   M. M. Lehman and Laszlo A. Belady. 1985. *Program Evolution: Processes of Software Change*. Academic Press Professional.

9   Linux GitHub repository, 'Pulse': https://github.com/torvalds/linux/pulse.

10 Puppet. 2017. State of DevOps 2017 (www.puppet.com/system/files/2017-state-of-devops-report.pdf).

11 Eric Ries. 2011. *The Lean Startup*. Portfolio Penguin.

12 Jonathan Kingham. 2019. Computer says no: facing up to the full implications of a digitised immigration system. Free Movement, 8 January (www.freemovement.org.uk/computer-says-no-digitised-immigration-system/).

13 Ministry of Civil Aviation (India). 2023. Digi Yatra app user base crosses the one million mark. Press release, 22 June (https://pib.gov.in/PressReleasePage.aspx?PRID=1934489). The Wire. 2024. DigiYatra: the promise and perils of india's new airport biometric system. *The Wire*, 8 January (https://thewire.in/tech/digiyatra-the-promise-and-perils-of-indias-new-airport-biometric-system).

14 Rosie Mears and Sophie Howes. 2023. You reap what you code. Child Poverty Action Group, 27 June (https://cpag.org.uk/sites/default/files/files/policypost/You_Reap_What_You_Code.pdf).

15 Rightsnet forum, 'Fantastic, interactive dummy UC claim form!': www.rightsnet.org.uk/forums/viewthread/18767/#88920.

16 Michael Buchanan. 2022. Pensions: millions receive wrong amount 'for decades'. *BBC News*, 17 June (www.bbc.co.uk/news/uk-61829661).

17 Douglas Adams. 2017 (1979). The Hitchhiker's Guide to the Galaxy. Pan.

18 Tom Bingham. 2011. *The Rule of Law*. Penguin.

19 'Screenshots/designs of the claimant user interface: a freedom of information request to Department for Work and Pensions', 6 September 2019: www.whatdotheyknow.com/request/ screenshotsdesigns_of_the_claima/response/1436957/attach/html/2/ FOI2019%2033077%20reply.pdf.html.

20 Linnean Society, 'Linnaean Collections': www.linnean.org/ research-collections/linnaean-collections.

21 Natural History Museum: https://data.nhm.ac.uk/dataset/collection-specimens/resource/05ff2255-c38a-40c9-b657-4ccb55ab2feb/ record/2221786/1715817600000.

22 '*Drosophila (sophophora) melanogaster* Meigen, 1830': www.gbif.org/ species/182839041.

23 B. Colwell. 2002. If you didn't test it, it doesn't work. *Computer* 35(5): 11–13 (https://ieeexplore.ieee.org/document/999770).

24 European Commission. 2023. The European Digital Identity Wallet architecture and reference framework. Policy document, 10 February (https://digital-strategy.ec.europa.eu/en/library/european-digital-identity-wallet-architecture-and-reference-framework).

25 'Eesti Rahvastikuregister': www.riha.ee/Infos%C3%BCsteemid/ Vaata/rr. Kristo Vaher. 2020. Next generation digital government architecture: version 1.0. Government Chief Information Officer, Estonia, March (https:// complexdiscovery.com/wp-content/uploads/2020/03/ Next-Generation-Digital-Government-Architecture-1.0.pdf).

26 'Design histories': https://design-histories.education.gov.uk/.

27 Flickr, 'The Help Forum: who decides interestingness?': www.flickr. com/help/forum/13790/.

28 Stuart J. Russell. 2019. *Human Compatible: Artificial Intelligence and the Problem of Control*. Allen Lane/Penguin Random House.

29 Bruce Schneier. 2023. AI to aid democracy. *Schneier on Security*, 26 April (www.schneier.com/blog/archives/2023/04/ai-to-aid-democracy.html).

30 Mustafa Suleyman and Ben Laurie. 2024. Trust, confidence and verifiable data audit. Google DeepMind, 14 May (https://deepmind. google/discover/blog/trust-confidence-and-verifiable-data-audit/). Philip Potter. 2015. Guaranteeing the integrity of a register. Technology in Government blog, 13 October (https://technology.blog. gov.uk/2015/10/13/guaranteeing-the-integrity-of-a-register/). Register Dynamics, 'Merkle tree-based logging': www.register-dynamics. co.uk/data-trusts/merkle-trees.

31 Apple App Store, 'Picture Insect: Spiders & Bugs': https://apps.apple.com/gb/app/picture-insect-spiders-bugs/id1461694973.

Chapter 10. Immunity to treachery

1 'Mozilla Stomps IE': https://home.snafu.de/tilman/mozilla/stomps.html.

2 Jon Swartz. 1997. Microsoft pulls prank / company takes browser war to Netscape's lawn. *SFGATE*, 2 October (www.sfgate.com/business/article/Microsoft-Pulls-Prank-Company-takes-browser-war-2803749.php).

3 US Department of Justice. 2015. US v. Microsoft: court's findings of fact. Report, 14 August (www.justice.gov/atr/us-v-microsoft-courts-findings-fact). R. A. Spinello. 2005. Competing fairly in the new economy: lessons from the browser wars. *Journal of Business Ethics* 57(4): 343–61 (https://doi.org/10.1007/s10551-005-1832-6).

4 Bill Gates. 1995. The internet tidal wave. Internal Microsoft email, 26 May (www.justice.gov/sites/default/files/atr/legacy/2006/03/03/20.pdf). *Wired*. 2010. May 26, 1995: Gates, Microsoft jump on 'internet tidal wave'. *Wired*, 26 May (www.wired.com/2010/05/0526bill-gates-internet-memo/).

5 Thomas A. Piraino, Jr. 2000. Identifying monopolists' illegal conduct under the Sherman Act. *New York University Law Review* 75(4): 809–92 (www.nyulawreview.org/wp-content/uploads/2018/08/NYULawReview-75-4-Piraino.pdf).

6 US District Court for the District of Columbia. 2002. Final judgement. November (www.justice.gov/atr/case-document/file/503541/dl).

7 Steve Lohr. 1999. Tiny software maker takes aim at Microsoft in court. *New York Times*, 31 May (https://archive.nytimes.com/www.nytimes.com/library/tech/99/05/biztech/articles/31soft.html).

8 Tech Law Journal, 'Summary: Bristol Technology v. Microsoft': www.techlawjournal.com/courts/bristol/Default.htm. *CNET*. 2002. Microsoft wins most of Bristol case. *CNET*, 6 February (www.cnet.com/tech/tech-industry/microsoft-wins-most-of-bristol-case/).

9 Lawrence Lessig. 2006. *Code: Version 2.0*. Basic Books (www.hachettebookgroup.com/titles/lawrence-lessig/code/9780786721962/).

10 Alan Murray. 2000. For policy makers, Microsoft suggests need to recast models. *Wall Street Journal*, 9 June (www.wsj.com/articles/SB960503996520272440).

11 Dacher Keltner. 2016. *The Power Paradox: How We Gain and Lose Influence*. Penguin.

12 Christoph Raetzsch, Gabriel Pereira, Lasse S. Vestergaard and Martin Brynskov. 2019. Weaving seams with data: conceptualizing city APIs as elements of infrastructures. *Big Data and Society* 6(1) (https://doi.org/10.1177/2053951719827619). Branden Hookway. 2014. *Interface*. MIT Press.

13 Lohr. 1999. Tiny software maker takes aim at Microsoft in court.

14 Apple. 2020. Apple and Google partner on Covid-19 contact tracing technology. Apple Newsroom, 10 April (www.apple.com/uk/newsroom/2020/04/apple-and-google-partner-on-covid-19-contact-tracing-technology/).

15 Google, 'Exposure notifications: helping fight Covid-19': www.google.com/covid19/exposurenotifications/.

16 Peter Wells, X.com post, 4 May 2020: https://x.com/peterkwells/status/1257565960883580929.

17 Marcel Salathé. 2023. Covid-19 digital contact tracing worked: heed the lessons for future pandemics. *Nature* 619(7968): 31–3 (https://doi.org/10.1038/d41586-023-02130-6). Rowland Manthorpe. 2024. Coronavirus: send virus alerts within 24 hours or risk second wave, scientist warns. *Sky News*, 8 May (https://news.sky.com/story/coronavirus-send-virus-alerts-within-24-hours-or-risk-second-wave-scientist-warns-11984908).

18 Jonathan Albright. 2020. The pandemic app ecosystem: investigating 493 Covid-related iOS apps across 98 countries. Medium, 15 December (https://d1gi.medium.com/the-pandemic-app-ecosystem-investigating-493-covid-related-ios-apps-across-98-countries-cdca305b99da).

19 Lindsay Clark. 2020. No surprise: Britain ditches central database model for virus contact-tracing apps in favour of Apple–Google API. *The Register*, 18 June (www.theregister.com/2020/06/18/uk_ditches_central_database_model/).

20 Michelle Kendall, Daphne Tsallis, Chris Wymant, Andrea Di Francia, Yakubu Balogun, Xavier Didelot, Luca Ferretti, *et al.* 2023. Epidemiological impacts of the NHS Covid-19 app in England and Wales throughout its first year. *Nature Communications* 14: paper 858 (https://doi.org/10.1038/s41467-023-36495-z).

21 Rowland Manthorpe, X.com post, 30 September 2020: https://x.com/rowlsmanthorpe/status/1311404110030217216.

22 Justin Elliott. 2019. Congress is about to ban the government from offering free online tax filing. Thank TurboTax. *ProPublica*, 9 April (www.propublica.org/article/congress-is-about-to-ban-the-government-from-offering-free-online-tax-filing-thank-turbotax). Justin Elliott and Paul Kiel. 2019. Inside TurboTax's 20-year fight to

stop Americans from filing their taxes for free. *ProPublica*, 17 October (www.propublica.org/article/inside-turbotax-20-year-fight-to-stop-americans-from-filing-their-taxes-for-free).

23 Justin Elliott and Lucas Waldron. 2019. Here's how TurboTax just tricked you into paying to file your taxes. *ProPublica*, 22 April (www.propublica.org/article/turbotax-just-tricked-you-into-paying-to-file-your-taxes). Justin Elliott. 2019. TurboTax deliberately hid its free file page from search engines. *ProPublica*, 26 April (www.propublica.org/article/turbotax-deliberately-hides-its-free-file-page-from-search-engines).

24 Internal Revenue Service, 'IRS Direct File pilot news': www.irs.gov/about-irs/strategic-plan/irs-direct-file-pilot-news. Internal Revenue Service. 2024. IRS makes Direct File a permanent option to file federal tax returns: expanded access for more taxpayers planned for the 2025 filing season. Press release, 30 May (www.irs.gov/newsroom/irs-makes-direct-file-a-permanent-option-to-file-federal-tax-returns-expanded-access-for-more-taxpayers-planned-for-the-2025-filing-season). Matt Bracken. 2024. House Republicans aim to end IRS's Direct File in 2025 appropriations bill. *FedScoop*, 6 June (https://fedscoop.com/house-republicans-irs-direct-file-cuts-appropriations-budget/).

25 Nicole Hartland. 2021. Parliament and the 1921 Poplar Rates Rebellion. Parliamentary Archives, 16 August (https://archives.blog.parliament.uk/2021/08/16/parliament-and-the-1921-poplar-rates-rebellion/). Steve Bloomfield. 2021. In the footsteps of George Lansbury: lost radical who led an East End rebellion. *The Guardian*, 31 July (www.theguardian.com/world/2021/jul/31/in-the-footsteps-of-george-lansbury-lost-radical-who-led-an-east-end-rebellion).

26 Richard Pope. 2019. Data infrastructure, APIs and open standards. In *Playbook: Government as a Platform*. Ash Center for Democratic Governance and Innovation, Harvard Kennedy School (https://ash.harvard.edu/wp-content/uploads/2024/02/293091_hvd_ash_gvmnt_as_platform_v2.pdf).

27 US General Services Administration, 'Products and platforms': http://web.archive.org/web/20190502001702/https://handbook.18f.gov/products-and-platforms/.

28 Dipartimento per la trasformazione digitale, 'Anagrafe Nazionale della Popolazione Residente': https://innovazione.gov.it/progetti/anagrafe-nazionale-della-popolazione-residente/.

29 Srinivas Kodali. 2019. Disenfranchised by Aadhaar: voter deletions in Telangana. *The Leaflet*, 17 March (https://theleaflet.in/disenfranchised-by-aadhaar-voter-deletions-in-telangana). *The Leaflet*. 2022. Supreme Court issues notice in petition alleging

arbitrary deletion of 46 lakh entries from electoral rolls in Andhra Pradesh, Telangana by Election Commission of India. *The Leaflet*, 14 December (https://theleaflet.in/supreme-court-issues-notice-in-petition-alleging-arbitrary-deletion-of-46-lakh-entries-from-electoral-rolls-in-andhra-pradesh-telangana-by-election-commission-of-india/).

30 Payal Arora. 2019. Benign dataveillance? Examining novel data-driven governance systems in India and China. *First Monday* 24(4) (https://doi.org/10.5210/fm.v24i4.9840).

31 Mark E. Warren (ed.). 1999. *Democracy and Trust*. Cambridge University Press (www.cambridge.org/core/books/democracy-and-trust/B46A44BAC288AACD3C79C55BEBDBC7C5).

32 Angeliki Kerasidou and Charalampia Kerasidou. 2023. Data-driven research and healthcare: public trust, data governance and the NHS. *BMC Medical Ethics* 24: paper 51 (https://doi.org/10.1186/s12910-023-00922-z).

33 Goods and Services Tax Network: www.gstn.org.

34 'The Aadhaar (Targeted Delivery of Dinancial and Other Subsidies, Benefits and Services) Act, 2016': https://uidai.gov.in/images/targeted_delivery_of_financial_and_other_subsidies_benefits_and_services_13072016.pdf.

35 Ministry of Civil Aviation (India). 2023. Digi Yatra app user base crosses the one million mark. Press release, 22 June (https://pib.gov.in/PressReleasePage.aspx?PRID=1934489). *The Wire*. 2024. DigiYatra: the promise and perils of india's new airport biometric system. *The Wire*, 8 January (https://thewire.in/tech/digiyatra-the-promise-and-perils-of-indias-new-airport-biometric-system).

36 Aheli Banerjee. 2023. Home Min can't deactivate Aadhaar without proper inquiry: Cal HC. *Times of India*, 7 October (https://timesofindia.indiatimes.com/city/kolkata/home-min-cant-deactivate-aadhaar-without-proper-inquiry-cal-hc/articleshow/104227962.cms).

37 Maansi Verma and Praavita Kashyap. 2023. In widening scope of Aadhaar, government is crossing red lines set by Supreme Court. *Indian Express*, 23 April (https://indianexpress.com/article/opinion/columns/in-widening-scope-of-aadhaar-government-is-crossing-red-lines-set-by-supreme-court-8571764/).

38 LocalGov Drupal, 'Web publishing for councils': https://localgovdrupal.org/.

39 Digital Transformation Agency, 'Your guide to the Digital Identity Legislation': https://web.archive.org/web/20240317160350/https://www.digitalidentity.gov.au/sites/default/files/2021-10/Your%20guide%20to%20the%20Digital%20Identity%20legislation.pdf.

40 Tom Loosemore. 2018. Making government as a platform real. Public Digital, 25 September (https://public.digital/pd-insights/blog/2018/09/making-government-as-a-platform-real).

41 Metropolitan Water Board. 1953. *London's Water Supply, 1903–1953: A Review of the Work of the Metropolitan Water Board.* Staples Press.

42 S. G. Hobson. 1905. Public control of electric power and transit. Report of the Committee of the Society appointed to consider the Control of Electrical Power and Transit. Fabian Society.

43 Trinity House, 'History of the Corporation': www.trinityhouse.co.uk/about-us/history-of-trinity-house/th500.

44 Based on the design by Stephen McCarthy for the Government Digital Service: www.loft27design.com. Also replicated here: https://public.digital/pd-insights/blog/2018/09/making-government-as-a-platform-real.

45 Jennifer Pahlka. 2023. *Recoding America.* Metropolitan Books.

46 Andrew Greenway, Ben Terrett, Mike Bracken and Tom Loosemore. 2021. *Digital Transformation at Scale: Why the Strategy Is Delivery,* 2nd edition. London Publishing Partnership.

Epilogue

1 Marie Hicks. 2018. *Programmed Inequality.* MIT Press. Hannah Fry (presenter). 2015. LEO the Electronic Office. Episode of *Computing Britain,* BBC Radio 4, 15 September (www.bbc.co.uk/programmes/b069rvb4).

2 Doug Chalmers. 2024. Post Office Horizon: an accountability failure. Committee on Standards in Public Life blog, 22 January (https://cspl.blog.gov.uk/2024/01/22/post-office-horizon-an-accountability-failure/). Kevin Peachey. 2021. Post Office scandal: what the Horizon saga is all about. *BBC News,* 23 April (www.bbc.co.uk/news/business-56718036).

3 'Le.taxi, le registre de disponibilité des taxis': https://le.taxi/.

4 Langdon Winner. 1980. Do artifacts have politics? *Daedalus* 109(1): 121–36 (www.jstor.org/stable/20024652).

5 Yu-Jie Chen, Ching-Fu Lin and Han-Wei Liu. 2018. 'Rule of trust': the power and perils of China's social credit megaproject. *Columbia Journal of Asian Law* 32(1): 1–36 (https://doi.org/10.7916/cjal.v32i1.3369).

6 European Commission. 2023. Road safety: Commission proposes updated requirements for driving licences and better cross-border enforcement of road traffic rules. Press release, 1 March (https://ec.europa.eu/commission/presscorner/detail/en/ip_23_1145).

Index

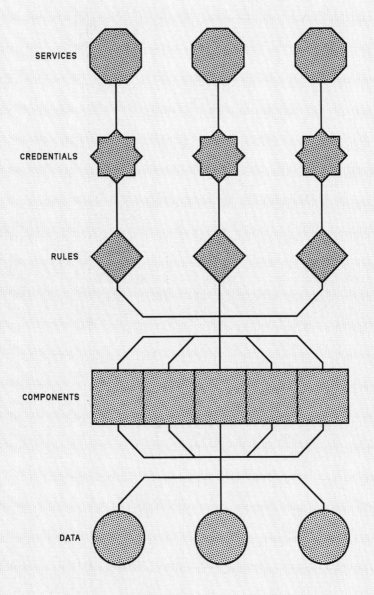

SERVICES

CREDENTIALS

RULES

COMPONENTS

DATA